Tsunami in Kerala, India: Long-Term Psychological Distress, Sense of Coherence,Social Support, and Coping in a Non-Industrialized Setting

Beiträge zur Pädagogischen und Rehabilitationspsychologie

Herausgegeben von Evelin Witruk

Band 5

PL ACADEMIC RESEARCH

Johanna Sophie von Lieres

Tsunami in Kerala, India: Long-Term Psychological Distress, Sense of Coherence, Social Support, and Coping in a Non-Industrialized Setting

PL ACADEMIC RESEARCH

Bibliographic Information published by the Deutsche Nationalbibliothek
The Deutsche Nationalbibliothek lists this publication in the Deutsche Nationalbibliografie; detailed bibliographic data is available in the internet at http://dnb.d-nb.de.

Zugl.: Leipzig, Univ., Diss., 2011

Cover Design:
© Olaf Gloeckler, Atelier Platen, Friedberg

Library of Congress Cataloging-in-Publication Data

Von Lieres, Johanna Sophie, 1978-
Tsunami in Kerala, India : long-term psychological distress, sense of coherence, social support, and coping in a non-industrialized setting / Johanna Sophie von Lieres.
pages cm. — (Beiträge zur Pädagogischen und Rehabilitations-psychologie, ISSN 1865-083X ; Bd. 5)
ISBN 978-3-631-62712-9
1. Post-traumatic stress disorder—India—Kerala. 2. Indian Ocean Tsunami, 2004. I. Title.
RC552.P67.V66 2013
616.85'210095483—dc23

2013007563

15
ISSN 1865-083X
ISBN 978-3-631-62712-9

Peter Lang Edition ist ein Imprint der Peter Lang GmbH.

Peter Lang – Frankfurt am Main · Bern · Bruxelles · New York · Oxford · Warszawa · Wien

www.peterlang.de

I dedicate this research work to Mata Amritanandamayi Devi (Amma).

“'My' house, ‘my’ life, ‘my’ belongings, ‘my’ children – all disappeared in the blink of an eye. Unexpected situations like this teach us that nothing is really ours.”

Amma
Tamil Nadu, February 12, 2005 (Mata Amritanandamayi Math, 2007, p.76)

Acknowledgements

I am very grateful for all that my supervisor Prof. Evelin Witruk has done for me. Her constant support, expert guidance, continuous encouragement, and very useful advice enabled me to pursue this PhD thesis work. I also would like to thank Dr. Frank Piontek for his invaluable assistance with my path analysis model.

Research for this dissertation would not have been possible without the generous financial support of the German Academic Exchange Service (Deutscher Akademischer Austauschdienst, DAAD).

Moreover, I would also like to thank all my family and friends from all over the world for their moral support throughout my dissertation project. A special thanks to my father who helped me with emotional, material, and practical support.

Moreover, I would like to thank all of the participants from the Alappad District for their time and effort. I owe special thanks also to the translators, without their help the study would not have been possible.

Finally, without Amma's grace and blessings, this dissertation project would not have been possible. It is She who gave me courage and strength. I humbly lay down this thesis paper at Amma's feet.

Preface

India was the third country severely affected by the tsunami on December 26, 2004, after Indonesia and Sri Lanka, with the overall death toll higher than 16,000. Kerala lost between 150 and 200 people. The present study investigated the long-term psychological consequences, and the role of sense of coherence, perceived social support, and coping after this natural disaster. The sample consisted of 407 inhabitants of the fishing villages along the coastline of Alappad, Kerala. Of these, 233 were affected by the tsunami and 174 were not affected. The level of traumatic stress was measured with the IES-R (Weiss & Marmar, 1996), general psychological distress with the BSI (Derogatis, 1993), sense of coherence with the SOC scale (Antonovsky, 1987), coping behavior with the Brief COPE (Carver, 1997), and perceived social support with the Social Support Questionnaire (SOZU-K-22) (Sommer & Fydrich, 1989).

The two groups were compared regarding their level of traumatization and their general psychological distress. In addition, it was assessed to what extent protective factors such as perceived social support, a sense of coherence, and specific coping strategies could have mitigated the effects of the disaster. It was revealed that tsunami victims were still suffering from symptoms of traumatic stress two and a half years after the tsunami and that they significantly showed greater psychological distress compared to the not affected control group. Social support, in particular, decreased the level of traumatic stress, and a strong sense of coherence mitigated psychological distress.

Contrary to studies in Western, industrialized settings, avoidance coping was found to be more effective in decreasing levels of traumatic stress than approach coping in this non-industrialized, collectivistic cultural setting.

Abbreviations

ACTH	Adrenocorticotropic hormone
AIMS	Amrita Institute of Medical Sciences
BSI	Brief Symptom Inventory
CCL	Coping Checklist
CRF	Corticotrophin-Releasing Factor
DSM	Diagnostic and Statistical Manual for Mental Disorders
ECA	Epidemiological Catchment Area
EDCF	East Coast Development Forum
F-SOZU	Fragebogen zur Sozialen Unterstützung (Social Support Questionnaire)
HARS	Hamilton Anxiety Rating Scale
HPA	Hypothalamic-Pituitary-Adrenal Axis
HSCL-25	Hopkins Symptom Checklist-25
IBT	Illinois Bell Telephone
ICD-10	International Classification of Diseases – 10
IES	Impact of Event Scale
IES-R	Impact of Event Scale – Revised
KMO	Kaiser-Meyer-Olkin Test
MAM	Mata Amritanandamayi Math
MMPI	Minnesota Multiphasic Personality Inventory
NGO	Non-Government-Organization
NIMH	National Institute of Mental Health
PAF	Principle Axis Factoring
PCA	Principle Component Analysis
PTSD	Post-Traumatic Stress Disorder
SCARF	Schizophrenia Research Foundation
SCL-90	Symptom Checklist-90
USGS	United States Geological Survey
WHO	World Health Organization

Contents

1 Introduction: Tsunami in Kerala ... 1

2 Description of Post-Traumatic Stress Disorder (PTSD) ... 7

2.1 Historical Development of the PTSD Construct ... 7
2.2 Definition of Trauma ... 8
2.3 Criteria for Diagnosis According to ICD-10 and DSM-IV ... 9
2.3.1 PTSD in the ICD-10 ... 10
2.3.2 PTSD in the DSM-IV ... 10
2.4 Cross-Cultural Validity of the PTSD Construct ... 15
2.5 Epidemiology ... 20
2.5.1 Community-Based Studies ... 21
2.5.2 Studies with High-Risk Groups ... 22
2.5.4 Studies in Non-Industrialized Settings ... 26
2.6 Prognosis and Duration of PTSD ... 28
2.7 Comorbidity with other Mental Disorders ... 30
3 Risk Factors and Predictors of PTSD ... 35

3.1 Socio-Demographic Factors ... 35
3.1.1 Gender ... 37
3.1.2 Age ... 39
3.1.3 Education ... 40
3.1.4 Earlier psychiatric disorders ... 41
3.1.5 Exposure ... 41
3.2 Predictors of PTSD: A Meta-Analysis ... 42
4 Protective Factors ... 43

4.1 Resilience, Hardiness, and Sense of Coherence in the Salutogenesis Model ... 43
4.1.1 Resilience ... 43
4.1.2 Hardiness ... 45
4.1.3 Sense of Coherence ... 46
4.2 Coping Strategies ... 53
4.2.1 Coping Efficacy ... 54
4.2.2 PTSD and Coping after Natural Disasters ... 56
4.2.3 Cultural Aspects of Coping ... 58
4.3 Social Support ... 62
4.3.1 Empirical Evidence ... 63
4.3.2 Social Support and the Family ... 66
4.3.3 Community Support and PTSD ... 67
4.3.4 Cultural Aspects of Social Support ... 67
5 Etiology of PTSD ... 69

5.1 Neurobiological Models ... 70
5.1.1 Neurotransmitters ... 70
5.1.2 Memory ... 72
5.2 Behavioral and Cognitive Models ... 73
5.2.1 Appraisals of the Trauma and/or its Consequences ... 73
5.2.2 The Characteristics of the Trauma Memory ... 74
5.2.3 Dysfunctional Behavior and Cognitive Processing Styles ... 75
6 Research Questions and Hypotheses ... 77
6.1 Research Questions ... 77
6.2 Hypotheses ... 83
7 Methods ... 85
7.1 Setting ... 85
7.1.1 Social Structure in an Indian Village ... 86
7.1.2 Religion and Ideology in Kerala ... 87
7.1.3 Cultural Beliefs about the Tsunami ... 88
7.2 Procedure ... 89
7.3 Participants ... 90
7.4 Instruments ... 93
7.4.1 Impact of Event Scale-Revised ... 93
7.4.2 Brief Symptom Inventory ... 99
7.4.3 Sense of Coherence Questionnaire ... 102
7.4.4 Brief COPE Scale ... 105
7.4.5 Social Support Questionnaire (Fragebogen zur Sozialen Unterstützung, F-SOZU; SOZU-K-22) ... 108
8 Results ... 111
8.1 Exploratory Factor Analyses ... 111
8.1.1 A Justification for the Choice of Factor Extraction and Rotation Methods ... 111
8.1.2 Approach to Calculation of Factor-Based Scores ... 113
8.1.3 Factor Analysis of the Impact of Event Scale – Revised (IES-R) ... 113
8.1.4 Factor Analysis of the Brief Symptom Inventory (BSI) ... 115
8.1.5 Sense of Coherence Scale (SOC Scale) ... 118
8.1.6 Brief COPE ... 120
8.1.7 Social Support Scale ... 121
8.2 Between-Group Comparisons ... 123
8.2.1 Effect Size and Test Power ... 125
8.3 Correlations ... 127
8.4 Regression Analyses ... 128
8.5 Comparing Demographic Factors in Relation to PTSD ... 133

8.6 Path Analysis Model 136
8.6.1 The Hypothesized Model 136
8.6.2 Data Analysis 137
8.6.3 Results of Path Analysis 138
9 Overview on the Confirmation of or Contradiction to the Hypotheses 146

10 Discussion 149

10.1 SOC and Traumatic Stress 150
10.2 Social Support and Traumatic Stress 150
10.3 Coping and PTSD 151
10.4 Risk and Protective Factors for Developing PTSD 155
10.5 Conclusions from the Path Analysis Model 157
10.6 Trauma from a Cultural Perspective 157
10.7 Critical Voices Against Using the PTSD Construct in a Non-Western Setting 159
10.8 Disaster Relief within a Collectivistic Context 162
10.9 Disaster Aid after the 2004 Tsunami: Lessons to be Learned 167
11 Conclusions 170

11.1 Implications for Practice 170
11.2 Limitations of the Study 171
11.3 Suggestions for Future Research 171
12 References 173

13 Appendix: Example of Efficient Disaster Relief Aid for Long- and Short-Term Rehabilitation 201

1 Introduction: Tsunami in Kerala

On December 24, 2004, an earthquake measuring 9.1 on the Richter scale occurred off the west coast of Northern Sumatra causing one of the most devastating tsunami waves ever recorded. It was the largest earthquake in the world since 1964. At least 231,000 people lost their lives and 1.7 million were rendered homeless (U.S. Geological Survey, USGS, 2008). Many people barely survived by running for their lives or by climbing onto rooftops (Schnibben, 2005). The under sea earthquake was caused by the Indian tectonic plate sliding under the Burma tectonic plate. The resulting ocean swells traveled as a wave 2,000 kilometers across the Indian Ocean. In this way, the tsunami waves caused destruction along the coastlines of 14 different countries, including Indonesia, Sri Lanka, Thailand, India, and even Africa (USGS, 2008).

India was the third country severely affected by the tsunami after Indonesia and Sri Lanka. The states affected were Tamil Nadu, Pondicherry, Andhra Pradesh, Kerala, and Andaman and Nicobar Islands. When the tsunami struck India's coastline at 8:45AM local time on December 26, 2004, the southeastern coast of Tamil Nadu and Andaman and Nicobar Islands were the worst hit areas (Arya, 2005).

The death toll in India was approximately 15,000 (Arya, 2005), or even higher than 16,000 with 10,749 dead and 5,640 missing on February 2, 2005 according to Mohanty (2005). Most of the Indians reported missing were from the Andaman and Nicobar Islands, so the death toll might actually be much higher. Immediately after the tsunami hit the Indian coastline, 730,000 individuals were forced to leave their homes. A total of 157,393 dwelling units were destroyed, without taking into account Andaman and Nicobar Islands (Government of India, Ministry of Home Affairs, 2005). Thus, even though some people were able to return to their houses, more than 400,000 people lost their homes (Tsunami Response Watch, 2005). One million people were altogether affected by the tsunami in India (International Federation of Red Cross and Red Crescent Societies, 2005), or even 1.3 million according to the Government of India, Ministry of Home Affairs (2005). In India, seventy-five percent of the people affected were women and children. A total of 83,788 boats were damaged or destroyed in India alone, thus depriving countless people of their source of livelihood (Government of India, Ministry of Home Affairs, 2005). The reconstruction costs were estimated to be more than 1.2 billion US dollars (The World Bank, 2005).

Table 1 shows the average scenario of tsunami devastation in the respective Indian states. Data relating to the Andaman and Nicobar areas are yet to be assessed, which is why they do not appear in the table (Mohanty, 2005; Government of India, Ministry of Home Affairs, 2005; World Health Organization, 2005).

Table 1

Tsunami Damage in India

Factor	Andhra Pradesh	Kerala	Tamil Nadu	Pondicherry	Total
Population affected	211,000	355,000	691,000	43,000	1,300,000
Area affected (Ha)	790	Unknown	2,487	790	> 4,067
Length of coast affected (Km)	985	250	1,000	25	2,260
Extent of penetration (Km)	0.5 - 2.0	1 - 2	1 - 1.5	0.30 - 3.0	
Reported height of tsunami (m)	5	3-5	7-10	10	
Villages affected	301	187	362	26	876
Dwelling units lost	1,557	11,832	91,037	6,403	110,829

In Kerala, the water level rose so dramatically that many houses were swept away or crumbled with the massive flooding. Many people lost their family members. In addition, many boats and fishing nets were destroyed and thus their means of earning a livelihood. Consequently, we can assume that the people being hit by the tsunami in the fishing villages of Kerala were traumatized by this natural disaster.

Kerala lost an estimated 200 people. On December 28, 2004, the death toll was already estimated at 156. The people most affected by the tsunami were the local fishermen. Eighty percent of the people affected by the tsunami came from fishing communities. Hundreds of fishermen lost their boats and other fishing equipment in the disaster. Moreover, the fishermen from Kerala experienced further problems, as the price of fish dropped by 30 percent in the weeks following the tsunami. People were afraid to eat fish because they thought the fish had been feeding on the dead bodies in the water (unpublished report by the students of Amrita Vishwa Vidyapeetham, Amritapuri, Kollam).

In Kerala, the Kollam district was the worst hit area with 131 casualties (Effect of 26 December 2004 Tsunami in Kerala Coast, 2005). According to the records of the Taluk Office in Karunagapally, Kollam District, Alappad Panchayat in the Kollam district was affected the most with 130 of the deaths and Azheekal being the worst affected village. A total of 1,443 people suffered from minor, and two from major injuries, and 2,909 houses were destroyed (August, 2007). A total of 450 people reported the loss of a family member and 1,500 lost their means of livelihood. Other districts in Kerala that were affected by the tsunami include Alapuzha, Ernakulam, Kannur, and Kozhikode (Report by the Taluk Office in Karunagapally, Kollam District).

The 2004 tsunami is just one example of a natural disaster. According to Linneweber and Lantermann (2006), natural disasters are taking place more and more frequently and causing increasing damage every year. During the past decade, the number of natural and technological disasters has dramatically risen. From 1994 to 1998, the number of reported disasters was on average 428 per year, but from 1999 to 2003 this has figure drastically increased by two-thirds to an average of 707 disasters annually. The greatest increase occurred in countries with low socio-economic power, where it has increased by 142 per cent. Both hydro-meteorological and geophysical disasters have become more common, becoming 68 and 62 percent more frequent, respectively, over the decade. Among natural disasters, floods are the most reported events in Africa, Asia, and Europe, while windstorms are most frequent in the Americas and Oceania. The reason why more and more people are being affected by disasters is due to a combination of factors: the rising number of disasters, drastic population increase in poorer parts of the world, and rapid and unplanned development, particularly in urban areas (International Federation of Red Cross and Red Crescent Societies, 2004). In a review paper on natural disasters in Asia, Kokai, Fujii, Shinfuku, and Edwards (2004) reported that of approximately the 3 billion people worldwide affected by disasters from 1967 to 1991, around 85 percent lived in Asia.

Numerous definitions of "disaster" can be found in the literature, such as the following:

> A disaster situation is the result of the interaction between a physical event (or a combination of several events) and communities or groups of people vulnerable to

> it, who do not have the resources to cope with the situation that ensues." (Revel, 1996, p.289).

According to Smith (1983, p.123), disasters are defined as "calamitous events, especially those occurring suddenly and causing great damage to property and hardships for human beings".

Such tragic events lead to a feeling of helplessness, being overwhelmed by uncontrollable events, and also to a shattering of basic assumptions about life. The victims of disasters experience a loss of personal security and the ability to predict the future. They see themselves as less capable of withstanding stressful situations. Moreover, the psychological reaction of most individuals after the disaster can be described as a reaction to extreme stress (Herzog, 2004). The oscillation between avoidance and intrusive memories of the event is described by Horowitz (1997) as a "stress response reaction". As people search for the meaning of such events, they often view them as "acts of God", punishing a community for misbehavior (Revel, 1996).

After a disaster has struck a community, it is not sufficient to provide the most essential physical needs, such as food, water, sanitation, shelter, and medical assistance. Relief organizations also need to establish psychological support, as well, as part of a rehabilitation project. The victims need to know that someone strong is in command and that they are now safe. At the same time, they need to be allowed to tell their story, as this enables them to express their distress. When survivors of natural disasters return to their homes, they need to readapt to their daily lives and they may develop new symptoms such as sleeping problems and poor concentration. Some people might even isolate themselves from former social contacts. Therefore, outreach programs for survivors should remain available for some time after the disaster (Revel, 1996).

Furthermore, as resources are scarce, external support is crucial in the aftermath of a disaster. Non-government organizations (NGOs) are the primary source for these extra resources. Their strength lies in their commitment, number of volunteers, sociocultural integration, and timely availability. Volunteers can attend to survivors and their families, listen to them, and respond to their questions and their needs. This reduces their acute level of stress and restores their self-confidence. NGOs also take care of the long-term-rehabilitation, which can take months or even years. They help the survivors establish a somewhat "normal" life and facilitate the communication between victims (Revel, 1996).

Since the Indian Government rejected help from foreign countries, it and Indian NGOs assumed the task of providing relief aid (Letukas & Barnshaw, 2008). One NGO, which provided immediate aid in the region where this study was conducted, is the Mata Amritanandamayi Math (MAM). In 2005, the United Nations conferred "Special Consultative Status" upon the organization with the UN Economic and Social Council. Starting immediately after the waves had struck the coastline of Kerala, where the headquarters of the MAM are situated, the MAM volunteers had started a massive tsunami relief and rehabilitation project. By September 2007, the MAM had constructed 4,500 tsunami-relief houses throughout India and Sri Lanka; distributed 700 fishing boats, boat engines and fishing nets; supplied food, clothing and medical care to hundreds of thousands; provided long term psychological counseling and wellness camps for children; established relief camps; built an escape bridge; and economically strengthened entire villages through the organization of cottage-industry cooperatives, education, and job training (Mata Amritanandamayi Math, 2007).

However, in spite of these massive relief efforts, the villagers had experienced a devastating traumatic event and the question arises, to what extent the people living along the coastline of India are still traumatized two and a half years after the tsunami. Immediately after the disaster, Western research on the consequences of the tsunami only focused on the psychological effects it had on tourists, their relatives and volunteers from abroad. Bronisch (2005) and Bronisch, Maragkos, Freyer, Müller-Cyran, Butollo, Weimbs, and Platiel (2006), for example, reported that German disaster management teams in Phuket and Kao Lak found German tourists with the following symptoms: dissociation, increased arousal, flashbacks, sleeping disorders, illusions, loss of appetite, grief reactions and suicidal ideation. In an article about her own activities as a helper, Zimmermann (2005) describes the great suffering that Swiss tourists in tsunami-affected areas had undergone. Even those German citizens who were not themselves present in the tsunami affected areas, but watched the events regularly on television, showed symptoms of anxiety and stress such as sleeping disorders and flashbacks (Knieper, 2006). But to what extent are the local people in the tsunami-affected areas traumatized?

As a matter of fact, extensive research has been done on long-term effects of natural disasters, however, most of the studies have been conducted in the Western hemisphere. Not many investigations have focused on answering the question to what extent the results of those studies are applicable in Non-Western cultures and can provide guidelines for effective relief aid and rehabilitation. Only few studies were conducted investigating the psychological sequelae of the tsunami, for example, Kumar, Murhekar, Hutin, Subramanian, Ramachandran, and Gupte, (2007) investigated the effects of the tsunami in Tamil Nadu. Therefore, the author wanted to explore the psychological consequences of the 2004 tsunami in South India.

The present study was conducted in certain fishing villages along the coastline of Kerala. This setting was chosen for several reasons: The population in

Kerala has the highest literacy rate in India, thus the participants were able to complete the questionnaires without difficulty. Along the coastline of Kerala, the author chose the worst hit areas and an area not affected by the tsunami that was nevertheless in close physical proximity to the affected area. Moreover, the participants had received some form of relief aid from the MAM and were thus willing to participate in the study. The majority of the villagers in the affected group, for example, had received new houses. In the not affected group, participants formed self-help-groups to start micro-businesses. Since the author is a member of the NGO MAM, she had the possibility to work together with leaders of the self-help-groups, and other villagers who supported her research by, for example, providing translators.

Taking into consideration cross-cultural aspects, a contribution shall be thus be made to exploring the long-term consequences of natural disasters. Why do some people suffer from PTSD, whereas others do not? Are certain protective factors also effective in non-Western countries? How do people cope with devastating flood disasters? What impact does the severity of the trauma, i.e. loss of one or more family members, destruction of one's house, or loss of ways to earn a livelihood, make on the level of impairment? Once it is known which protective factors mitigate the traumatizing effects of natural disasters, this information could influence future long-term rehabilitation projects for disaster victims.

2 Description of Post-Traumatic Stress Disorder (PTSD)

When a life-threatening event occurs, it is normal to react with distress, anxiety, and fear. This reaction enables people to survive and these emotions enhance the memory of the traumatic event, so that they can recognize and avoid similar situations in the future. For some individuals, however, this natural reaction to a traumatic situation becomes uncontrollable and exaggerated, and they develop symptoms of PTSD. The nature and severity of the trauma as well as personal characteristics determine the severity and duration of the symptoms (Breslau, 2001b).

When one's life is in danger, one reacts to this situation of extreme stress by, for example, denying what has happened or giving the impression of losing contact with reality. Many people cannot quickly integrate what has happened and refuse to accept reality. They may feel numb and focus on insignificant details. Others feel guilty and in some way responsible for the tragedy. Fear of the reoccurrence of the event is also common. These are normal reactions to abnormal situations that help people survive (Revel, 1996).

2.1 Historical Development of the PTSD Construct

Psychiatrists' and psychologists' understanding of PTSD has undergone a series of changes throughout the last 150 years. The diverse social, biological, and psychological processes associated with it were initially believed to have only physical causes.

In 1866, the term "railway spine" was first used to indicate a symptom group associated with events such as railway collisions, producing shock, fright and emotional disturbance. The opinion of the majority was that the jolts and shakes experienced on these occasions were sufficient to cause neurological damage. The novelty about this syndrome was that no significant external injury could be seen (Young, 2000).

A few years later in 1871, the phenomenon of soldiers being affected by the events of the American Civil War was described in the scientific literature as "Da Costa syndrome", "irritable heart" (Da Costa, 1871), "effort-syndrome", or "neurocirculatory asthenia". These terms refer mainly to the physical symptoms that the soldiers developed as a reaction to the psychological traumatization of the Civil War.

In the late 19th century, terms such as "hysteria" and "traumatic neuroses" were common in Europe. A "traumatic neurosis" was said to be a mental disorder caused by a physical as well as psychological trauma such as an illness or the like (Berger & Van Calker, 2004). Freud differentiated a traumatic neurosis from anxiety and depressive neuroses and postulated that the term "traumatic" refers to an experience that within a short period of time presents the mind with an increase in stimulus too powerful to deal with in a normal manner. During the First World War, the traumatized soldiers were supposedly afflicted by "shell-shock", resulting from microhemorrhages caused by the explosion of shells, or others believed that they simply refused to return to the war front due to their lack of courage. Consequently, they were considered as "moral invalids" and treated with disciplinary therapy, which involved severe electric shocks or other methods to inflict pain. This view and these treatment methods continued into the Second World War (McFarlane, 2000).

Survivors of the Holocaust and concentration camps of the Nazi-regime suffered from the alleged "survivor syndrome", which manifested in symptoms such as psycho-physiological exhaustibility, as well as depressive and anxious symptoms that are still visible in the victims today (Berger & Van Calker, 2004).

Finally, in 1980, it was recognized that survivors of traumatic events generally endure specific and painful psychological consequences of the event. The term "post-traumatic stress disorder" was introduced in the DSM-III (American Psychiatric Association, 1980), as the diagnostic criteria of the Diagnostic and Statistical Manual of Mental Disorders (DSM). This term carried over into the DSM-III-R (American Psychiatric Association, 1987) and the DSM-IV (American Psychiatric Association, 1994), and can be found in the chapter for anxiety disorders (Berger & Van Calker, 2004). In the International Classification of Diseases (ICD-10), the PTSD is not a subgroup of the anxiety disorders but of the category F4, "Neurotic, Stress and Somatoform Disorders". The anxiety disorders are also a subgroup of this category (World Health Organization, 1992).

2.2 Definition of Trauma

A traumatic event that is defined as a "trauma" and that precedes PTSD, is described by the World Health Organization (WHO) (1992) as:

> A stressful event or situation (either short- or long-lasting) of an exceptionally threatening or catastrophic nature, which is likely to cause pervasive distress in almost everyone (e.g. natural- or man-made disaster, combat, serious accident, witnessing the violent death of others, or being the victim of torture, terrorism, rape, or other crime) (World Health Organization, 1992, p.147).

In 1987, the American Psychiatric Association defined a traumatic event as one "that is outside the range of usual human experience and that would be markedly distressing to almost anyone" (p.250). However, in the most recent Diagnostic and Statistical Manual of Mental Disorders (DSM-IV) of the American Psychiatric Association the term "trauma" is defined in a more narrow sense. An event needs to meet two criteria to be called a trauma: (1) "a person experienced, witnessed, or was confronted with an event or events that involved actual or threatened death or serious injury, or a threat to the integrity of self or others", and (2) "the person's response involved intense fear, helplessness or horror (in children, disorganized or agitated behavior)" (American Psychiatric Association,1994).

2.3 Criteria for Diagnosis According to ICD-10 and DSM-IV

Certain similarities and differences exist between the diagnostic criteria of DSM-IV and ICD-10. The most important difference between the two systems is the emphasis placed on symptoms of "emotional numbing". Three symptoms of emotional numbing need to be present according to the DSM-IV for the diagnosis of PTSD to be given. In the ICD-10 however, emotional numbing is not regarded as necessary to make the diagnosis, although the ICD-10 does recognize it as a frequent accompaniment to PTSD. The ICD-10 prefers that only one diagnosis be given to a patient, whereas the DSM-IV system encourages the making of multiple diagnoses, thereby increasing the amount of comorbidity reported. (Yule, 2001b; Shalev, 2001).

Moreover, there are differences in the time factor. For example, in the ICD-10, there needs to be an obvious time connection between the stressor and the *acute stress reaction*, which usually subsides within hours or days, whereas the *acute stress disorder* of the DSM-IV can only be diagnosed two days after the stressful event and it does not need to manifest immediately after the event. The acute stress disorder can begin up to four weeks after the event and its symptoms can persist for four weeks. If a certain pattern of symptoms continues on after the initial four weeks, it is then called *post-traumatic stress disorder*. If the symptoms last for less than three months, the PTSD is "acute", and if they persist for more than three months, the PTSD is labeled "chronic". If the symptoms begin six months or more after the traumatic event, the patients are said to have "delayed-onset" PTSD (Krüsmann & Müller-Cyran, 2005; Yule, 2001b). According to Krüsmann and Müller-Cyran (2005), the different concepts regarding the onset and duration of the symptoms are not a contradiction, but move along a time continuum. In the immediate aftermath of a trauma, almost everyone experiences a series of severe stress reactions that is considered normal in these circumstances.

2.3.1 PTSD in the ICD-10

In the ICD-10, the post-traumatic stress disorder appears in Chapter F43 as a "Reaction to severe stress, and adjustment disorders". It is the only category, which not only describes the symptomology but also indicates causative influences for the disorders (acute severe stress or continued trauma). According to ICD-10, PTSD needs to be distinguished from an acute stress reaction and the adjustment disorder. An exceptionally stressful life event produces an acute stress reaction, which includes a "state of 'daze', with some constrictions in the field of consciousness and narrowing of attention, inability to comprehend stimuli and disorientation. This state may be followed either by further withdrawal from the surrounding situation (to the extent of a dissociative stupor – see F44.2), or by agitation and over-activity (flight reaction or fugue)" (World Health Organization, 1992, pp. 146-147). However, the symptoms usually subside within hours or days. A significant life change leading to continued unpleasant circumstances can also result in an adjustment disorder, manifesting in "depressed mood, anxiety, worry (or a mixture of these), feeling of inability to cope, plan ahead, or continue the present situation, and some degree of disability in the performance of daily routine" (World Health Organization, 1992, p.149). The onset is within one month of the occurrence of the event or life change, and the duration does not exceed six months.

The diagnostic guidelines for F43.1 post-traumatic stress disorder are as follows:

> This disorder should not generally be diagnosed unless there is evidence that it arose within 6 months of a traumatic event of exceptional severity. A 'probable' diagnosis might still be possible if the delay between the event and the onset was longer than 6 months, provided that the clinical manifestations are typical and no alternative identification of the disorder (e.g. as anxiety or obsessive-compulsive disorder or depressive episode) is plausible. In addition to evidence of trauma, there must be a repetitive, intrusive recollection or reenactment of the event in memories, daytime imagery, or dreams. Conspicuous emotional detachment, numbing of feeling, and avoidance of stimuli that might arouse recollection of the trauma are often present but are not essential for the diagnosis. The autonomic disturbances, mood disorder, and behavioral abnormalities all contribute to the diagnosis but are not of prime importance. The late chronic sequelae of devastating stress, i.e. those manifest decades after the stressful experience, should be classified under F62.0" (World Heath Organization, 1992, pp.148-149).

2.3.2 PTSD in the DSM-IV

In the DSM-IV, PTSD is classified as an anxiety disorder. The diagnosis is given independent of the kind of traumatic event. Therefore, the description of

PTSD as a psychological long-term effect of an extreme stressor is described as a final common pathway of diverse adverse events (Maercker & Ehlert, 2001). The following main symptoms need to be present for the diagnosis of PTSD: (1) intrusion, (2) avoidance of stimuli associated with the trauma, and (3) increased arousal in general.

Intrusion

When someone suffers from intrusion, he or she has recurrent and intrusive distressing recollections of the event. These memories are often not part of the "language memory system", but consist of fragmented reliving of the traumatic experience. A sound, a smell or an image can come into mind, or the person may have the impression of being back in the traumatic situation – which is called a "flashback". Nightmares of the event can also occur, or strong psychophysical reactions to internal or external cues that symbolize or resemble an aspect of the traumatic event (Krüsman & Müller-Cyran, 2005). Intrusions can also take the form of intense emotions, such as panic or rage, or somatic sensations, interpersonal reenactments, or character styles. PTSD patients are not capable of integrating traumatic experiences with other life events, and their traumatic memories are not coherent stories (van der Kolk, 2000).

Avoidance

PTSD patients avoid internal and external cues reminiscent of the trauma. These can be thoughts, emotions, or discussions about the event. Traumatized individuals organize their entire life around avoiding the emotions that the intrusions evoke. To feel nothing seems to be better than to feel irritable and upset. Apart from keeping away from any reminders, avoidance can also take the form of substance abuse to numb emotional states or utilizing dissociation to keep traumatic experiences from entering an individual's conscious awareness. Since people avoid certain places, situations or activities, this can lead to social withdrawal. Individuals suffering from PTSD might lose interest in activities they used to enjoy and feel estranged from other people. Avoidance also includes an incomplete memory of the trauma and emotional numbing. Usually, a numbing of responsiveness to a wide range of emotional aspects of life takes place (van der Kolk, 2000; Krüsman & Müller-Cyran, 2005). Avoidance can also be considered as a reaction to the intrusion symptoms. People suffering from intrusive memories try to avoid these by avoiding all activities, situations, and emotions that are associated with the traumatic event (Maercker & Ehlert, 2001). Intrusive experiences trigger an opposing process of cognitive and emotional denial that represents the defensive phase of adjustment. The function of denial is to ward off the painful effects and memories related to the trauma. A traumatized individual is said to shift back and forth between generalized unresponsiveness (numbing) and intrusions until resolution of the trauma occurs (Horowitz, 1986). According to Litz, Orsillo, Kaloupek, and Weathers (2000), individuals suffering from PTSD are not generally numb and unresponsive. Rather, what has earlier been labeled as "emotional numbing" is best described as a deficit in emotional processing that stems from episodes of hyperemotionality caused by reminders of the trauma. Therefore, when in a state of acute stress triggered by trauma cues, PTSD patients are hypothesized to experience less intense positive feelings and to be more reactive to negative cues. In a study conducted by these authors with Vietnam War veterans, those veterans suffering from PTSD showed less expressive behaviors that would otherwise signal a positive response to positively valenced images after being exposed to reminders of the trauma, whereas veterans without PTSD showed normal emotional responses. However, the PTSD group had greater heart rates when exposed to all emotional stimuli. The difference between the two groups was in their emotional expression consistent with pleasant feelings and not in their emotional experience.

Increased Arousal

Even though emotional numbing occurs, the bodies of traumatized individuals' still react to certain physical and emotional stimuli as if there was a continued threat. After a traumatic event has happened, the level of arousal can be altered permanently (hyperarousal) and people with PTSD may suffer from sleep disturbances, lack of concentration, hypervigilance, exaggerated startle response, and restlessness. They also suffer from conditioned autonomic arousal to trauma-related stimuli as well as extreme physiological arousal in response to a wide variety of other stimuli. When they experience intense negative emotions as a reaction even to minor stimuli, they either overreact or threaten others or they shut down and freeze. They do not process information correctly, as every stimulus is perceived as a potential indicator of danger and their range of perception is significantly narrowed down. They view the world as an unsafe place. Moreover, people suffering from PTSD can no longer rely on autonomic arousal alerting them to danger, as their autonomic nervous system constantly fires warning signals. Bodily sensations no longer function as signals of emotional states, and stop serving as guides for action. The individual's own physical sensations become a source of fear (van der Kolk, 2000; Krüsmann & Müller-Cyran, 2005).

DSM-IV criteria for the diagnosis of post-traumatic stress disorder (PTSD)

A The person has been exposed to a traumatic event in which both the following were present:

1. The person experienced, witnessed, or was confronted with an event or events that involved actual or threatened death or serious injury, or a threat to the physical integrity of self or others.
2. The person's response involved fear, helplessness, or horror. (Note: in children, this may be expressed as disorganized or agitated behavior.)

B The traumatic event is persistently reexperienced in one (or more) of the following ways:

1. Recurrent and intrusive distressing recollection of the event, including images, thoughts, or perceptions. (Note: In young children, repetitive play may occur in which themes or aspects of the trauma are expressed.)
2. Recurrent distressing dreams of the event. (Note: Children may have frightening dreams without recognizable content.)
3. Acting or feeling as if the traumatic event were recurring (includes a sense of reliving the experience, illusions, hallucinations, and dissociative flashback episodes, including those that occur upon awakening or when intoxicated). (Note: in young children, trauma-specific reenactment may occur.)
4. Intense psychological distress at exposure to internal or external cues that symbolize or resemble an aspect of the traumatic event.
5. Physiologic reactivity upon exposure to internal or external cues that symbolize or resemble an aspect of the traumatic event.

C Persistent avoidance of stimuli associated with the trauma and numbing of general responsiveness (not present before the trauma), as indicated by three (or more) of the following:

1. Efforts to avoid thoughts, feelings, or conversations associated with the trauma.
2. Efforts to avoid activities, places, or people that arouse recollections of the trauma.
3. Inability to recall an important aspect of the trauma.
4. Markedly diminished interest or participation in significant activities.
5. Feelings of detachment or estrangement from others.
6. Restricted range of affect (e.g., unable to have loving feelings)
7. Sense of foreshortened future (e.g., does not expect to have a career, marriage, children, or a normal life span.)

D Persistent symptoms of increased arousal (not present before the trauma), as indicated by two (or more) of the following:
1. Difficulty falling or staying asleep.
2. Irritability or outbursts of anger.
3. Difficulty concentrating.
4. Hypervigilance.
5. Exaggerated startle response.

E Duration of the disturbance (symptoms in criteria B, C, and D) is more than one month.

The disturbance causes clinically significant distress or impairment in social, occupational, or other important areas of functioning.

Specify if:
Acute: If duration of symptoms is less than three months.
Chronic: If duration of symptoms is three months or more.

Specify if:
With delayed onset: If onset of symptoms is at least six months after the stressor.
(American Psychiatric Association, 1994)

The criteria according to DSM-IV – especially the avoidance-criteria – are quite strict. For this reason, a partial PTSD was introduced, which occurs very frequently and is as stable as the full PTSD (Shalev, 2000). It is also suggested that only two symptoms be required to meet the "C" criteria until more evidence suggests otherwise. The current requirement of "three symptoms from C" may artificially reduce the number of people who receive the diagnosis (Green, 1993).

Other authors have also criticized the requirement of the "C" criteria. For example, McFarlane (1990) states that by focusing on the withdrawal from social relationships and on avoidance, an individual who is premorbidly an introverted person is more likely to show the DSM group "C" criteria for PTSD. Consequently, an individual who does not respond to the intrusive memories by withdrawal and avoidance will not be given a diagnosis of PTSD even though he or she is significantly impaired.

The consequences of a traumatic event can also manifest in more ways than in PTSD symptoms. Other areas of an individual's life can be also affected, as well. For example, marriages or any other human relationships can suffer. Career development can be hampered and people may lose their personal interests and creativity. All in all, life can lose its joy and flavor (Krüsmann & Müller-Cyran, 2005).

One set of characteristic behaviors not mentioned in the DSM-IV is the compulsive reexposure to the trauma. Traumatized individuals often expose themselves to situations that remind them of the trauma. For example, combat soldiers become policemen, abused women marry men who mistreat them, or sexually abused children become prostitutes. When the traumatic situation is reenacted, the individual can either play the role of the victimizer or victim. So far, this phenomenon has not been sufficiently explained (van der Kolk, 2000).

Children and adolescents show symptoms according to their age such as anxious clinging, regression to an earlier stage of development, withdrawal, persisting fears and sleep disturbances (Gordon, Farberow & Maida, 1999).

Some authors have also voiced their critical opinions about the DSM Criteria. (McNally, 2004). For example, in the beginning, psychiatrists opposed to the inclusion of the diagnosis in DSM-III argued that the psychological problems of trauma survivors were already covered by combinations of existing diagnoses, such as phobias, depression, or personality disorders. Selected symptoms exhibited by people suffering from multiple disorders were grouped together and then attributed to a traumatic event. Moreover, the fact that the movement to include PTSD in the DSM diagnostic system originated from Vietnam veterans' advocacy groups working with anti-war psychiatrists arouses concerns that the PTSD construct was more of a social or political construct, rather than a phenomenon discovered in nature. McNally (2004) also criticizes that in the DSM-IV, the concept of a traumatic stressor has broadened to such an extent that today the vast majority of adult American citizens have been exposed to PTSD-qualifying events. This contradicts the trauma definition of the late 1970s and early 1980s, when a trauma had to be a catastrophic event falling outside the perimeter of everyday experience. In the DSM-IV, the subjective experience and emotional reactions during the event of the trauma survivor are crucial for the diagnose. He or she has to experience "intense fear, helplessness, or horror". Therefore, if anyone dissociates during the trauma and thus blunts out emotional experience, that person cannot be diagnosed as having PTSD. However, research has shown peritraumatic dissociation to be among the best predictors of subsequent PTSD (Ozer, Best, Lipsey, & Weiss, 2003). Finally, "PTSD is not 'discovered' in nature, but co-created via the interaction of psychobiology and the cultural context of classification" (McNally, 2004, p.11).

2.4 Cross-Cultural Validity of the PTSD Construct

Since this study has been conducted in a non-Western, non-industrialized setting, the question arises whether PTSD is a cultural construct only applicable in Western, industrialized countries or whether PTSD symptomology is biologically based and therefore universal. Most often, treatment methods and models for traumatic stress have been derived from studies of samples taken from industrialized countries and yet they are increasingly applied to diverse cultural populations (Yeomans & Forman, 2009). However, human response to traumatic events could be different in different settings. The reliability and validity of the PTSD measurement, what events diverse groups consider as traumatic, how they respond to trauma, and other factors still need to be investigated (Marsella, Friedman, & Spain, 1996). Several authors have attempted to answer this question by conducting studies in non-industrialized countries.

For example, Yeomans, Herbert, and Forman (2008) conducted a study to describe traumatic stress reactions in a Burundian sample (Africa). They investigated the influences of open-ended questions versus standardized measures and also the influence of psychoeducation – as a process of acculturation – on PTSD symptoms. Content analysis of open-ended questions revealed frequent material complaints (85.9% of the sample), depression (41.1%), PTSD (39.8%), anxiety (34.6%), somatic/medical (25.7%), and anger (11.5%), whereas standardized measures showed traumatic stress manifesting more as somatization, anxiety, and depression but less as specific PTSD symptoms. Psychoeducation – which exposed the respondents to Western ideas about trauma - was predictive of more severe PTSD symptoms. This study shows how solicitation methods and cultural factors may shape posttraumatic stress presentation. The low level of PTSD measured quantitatively could result from a variety of factors: it might indicate the lack of validity of standardized measures, the symptoms might have remitted over time, or the respondents might have underreported the symptoms. According to the authors, the possibility still remains that Western trauma models capture a traumatic stress response that is not yet fully articulated by indigenous groups. However, the universality of PTSD symptoms strongly depends on the solicitation methods and cultural ideas inherent in contemporary Western trauma discourse.

Another study in a non-industrialized setting is a survey of civilian war victims Freetown, Sierra Leone, in Western Africa (Raymond, 2000). The author measured the consequences of traumatic stress with the Impact of Event Scale (IES) (Horowitz, Wilner, & Alvarez, 1979), and reported a 99 percent PTSD prevalence rate. Even though the IES has not yet been validated for Western Africa and the applied cut-off scores are mostly based on European samples, it has been used worldwide and consistent structures have been found across samples and situations. In this sample, responses to traumatic stress were also found to be strongly associated with somatization. Since the onset of the violence in Sierra Leone, eighty-five percent of the respondents perceived their health to have worsened. Non-specific physical complaints were reported by 78 percent and 42 percent had visited the health post or clinic at least twice in the preceding four weeks.

However, according to Yeomans et al. (2008), one needs to be careful when interpreting such extraordinary results as it is necessary to consider the quality of translation, validity of the measures, broadness of the symptom assessment, and social influences (social desirability, power differential, overevaluation of Western culture).

The cross-cultural applicability of the PTSD concept was also investigated by assessing symptom severity and levels of comorbid psychopathology in adolescents from the United States and Russia in a self-report survey (Ruchkin, Schwab-Stone, Jones, Cicchetti, Koposov, & Vermieren, 2005). In both countries, the levels of all three clusters of PTSD symptoms (re-experiencing, avoidance, and arousal) as well as comorbid psychopathology increased as the level of experienced post-traumatic stress increased. Moreover, no differences in significant interaction effects for symptom levels were found between countries. These results imply that PTSD symptoms and their comorbidity with other mental health problems are not culture-bound and that the psychological consequences of traumatic events follow similar dynamics across cultures.

When Sack, Seeley, and Clarke (1997) attempted to replicate the factor structure of the Diagnostic Interview for Children and Adolescents with a sample of Khmer (Cambodian) refugees, they succeeded and found the same factor structure as in industrialized samples. The following four factors were found: arousal, avoidance, intrusion, and numbing, accounting for 58.5 percent of the variance.

Likewise, the same factor structure of the 13-item IES for children was found when used in a sample of Bosnian and British children aged 9 to 14 (Smith, Perrin, Dyregrov, & Yule, 2003). The 13-item IES for children is an extended version of the short 8-item version with an additional five arousal items. After Varimax rotation, in both samples, two clear factors corresponding to intrusion and avoidance items emerged, accounting for 54.6 percent of the variance. The conclusion drawn from these results is that, with some discrepancies, post-traumatic stress reactions in children are more similar than different across cultures and PTSD is not a culture-bound symptom.

Horowitz's IES was evaluated in a meta-analysis by Sundin, Mardi, and Horowitz (2003), and the results indicated that gender and cultural differences were only weak predictors for intrusive and avoidance symptoms, whereas the type of event and time elapsed since the event were strong predictors. These data provide evidence for the value of the IES as a measure of stress reactions in a number of different populations.

Even though these results indicate the cross-cultural validity of the PTSD construct, it remains a controversial issue. Some authors suggest that the DSM-IV criteria do not cover the full spectrum of the experience of traumatic stress of indigenous people. A broader range of responses to traumatic situations should be included. These authors chose less structured interview techniques in order to capture more fully the breadth of possible symptoms (Ruchkin et al., 2005). Even though these approaches are not well-suited to assess prevalence rates, they can better characterize the local manifestation of traumatic stress (Yeomans & Forman, 2009).

Baron (2002) conducted qualitative analyses with Sudanese refugees in northern Uganda and found a consistent pattern of symptoms that is broader than the usual definition of PTSD: anxiety, numerous somatic complaints, standard depressive symptoms, estrangement from friends and family, and loss of motivation to care for family and self. Moreover, refugees consistently focused more on survival concerns, such as lack of food, poor health care, or threat of violence, rather than on the traumatic event and its psychological consequences.

Symptom Variation

Other results have shown considerably more variability in the manifestation of specific post-traumatic stress symptoms among different cultures. The biological and psychological processes of human response to stress may be universal, but the specific PTSD response may vary across cultures, especially with regard to the re-experiencing and avoidance dimensions (Marsella & Christopher, 2004). Therefore, it can be said that the PTSD construct has both culture-bound and universal dimensions. For example, Marsella, Friedman, Gerrity, and Scurfield (1996) suggested that intrusive symptoms might be universal, whereas avoidance/numbing symptoms may be more influenced by culture. Other authors also came to the same conclusion (Dyregrov, Gupta, Gjestad, & Mukanoheli, 2000; McCall & Resick, 2003). Marsella et al. go so far as to suggest that PTSD levels might be higher in cultures in which avoidance/numbing symptoms are more prevalent (Marsella, Friedman, Gerrity, & Scurfield, 1996). In any case, these authors warn against using PTSD as a unidimensional construct and suggest using a broader spectrum of assessment. Moreover, in many non-western cultures, the control of one's emotions is socially desirable and open displays of anger are not permitted. Thus, the culture as a whole encourages emotional numbing and avoidance (Zur, 1996). There is also a complex association between the expression of distress and the occurrence of somatic symptoms in non-Western cultures. It has been claimed that Asians display their distress in a more somatic way than Westerners (Kinzie, Boehnlein,

Leung, Moore, Riley, & Smith, 1990). In a study of Tibetan refugees in North India, Terheggen, Stroebe, and Kleber (2001) found that participants on average scored higher on items phrased in a somatic way. For non-Western people, referring to somatic symptoms and using metaphors is a socially acceptable way of expressing emotional distress (Zur, 1996).

Moreover, in many non-Western cultures, fatalism is quite common. Belief in the arbitrariness of fate is an important factor in the normalization of adverse events for example among the Quiche Mayans. It protects one from the ravages of random circumstance. Thus, instead of suffering great psychological distress over, for example, losing the entire male presence in a family, there is self-denial, self-effacement and resignation to the fate of enormous loss. They believe. what befalls a person depends on his or her fate (Zur, 1996).

This observed variability in symptom manifestation could also be attributed to methodological and sampling differences rather than cultural differences. However, there is still no definite answer to the question of whether the results imply PTSD being a cultural-bound disorder or the inevitable variability of a universal construct. Studies have also investigated the biological markers of PTSD to demonstrate the universality of the construct. In a study with Armenian youth, those with greater earthquake exposure and subsequent PTSD symptoms had more depressed cortisol levels than non-exposed youth, thus indicating higher levels of PTSD (see 5.1) (Goenjian, Yehuda, Pynoos, Steinberg, Tashjian, Yang et al., 1996). The findings are congruent with those previously described in adults with chronic PTSD. Persistent intrusion symptoms may constitute episodes of distress and evoke repeated physiological stress responses, which, over time, alter hypothalamic-pituitary-adrenal (HPA) axis function. These findings suggest that certain biological markers are associated with PTSD across cultures, but conclusions about causality cannot be made (Yeomans & Forman, 2009). Even though traumatic stress reactions tend to have a universal biological underpinning, the vulnerability to trauma, symptom manifestation, and response to treatment are quite likely to be subject to contextual and cultural factors (Marsella, Friedman, & Spain, 1996).

Moreover, another conceptual issue concerns the nature of traumatic stressors. These might not be the same in different cultural settings. A standard checklist could underestimate the prevalence of extreme stressors if the list was not moderated according to the cultural values of a particular population. For example, in a study with Tibetan refugees living in North India, the destruction of temples, monasteries, and other religious signs was found to be an extremely stressful and frequently experienced traumatic event. It is interesting to note that the Impact of Event Scale (Horowitz, et al. 1979) could not be used in this sample. The scale could not be adequately translated into Tibetan, as the items became almost identical and thus redundant. Rather, the participants were only asked about the occurrence of intrusions and avoidance of thoughts related to the traumatic event and grouped into two groups: those who had no intrusion-avoidance and those who had intrusion or avoidance symptoms or both. Participants in the latter group had significantly higher anxiety and depression scores and non-significantly higher levels of somatic complaints (Terheggen et al., 2001).

As mentioned above, in cross-cultural research, methodology significantly influences the outcome. Both quantitative and qualitative methods have their advantages and disadvantages. Quantitative methods impose preestablished constructs and methods from a cultural perspective foreign to the sample under study. In contrast, qualitative methods using open-ended questions aim at collecting indigenous models of traumatic stress and assessing a broad spectrum of possible symptom manifestations. However, in qualitative methods, symptoms that are considered too private or insignificant to disclose are easily overlooked. Moreover, opportunities for statistical analyses are limited. Quantitative analyses are easier in questionnaire-based methods (Yeomans & Forman, 2009).

In conclusion, much of the literature indicates that in its different variations around the world, PTSD appears to be universally evident, as stated by Marsella and Christopher (2004).

2.5 Epidemiology

Most people experience at least one traumatic event during their lifetime, and most are able to deal with it and move on. However, for some people the exposure to such an event marks the beginning of a period of suffering and distress, as they develop the symptoms of PTSD. Although this period can be relatively short-lived, for some people it can remain unremitted for many months or even years (Breslau, 2001a).

Studies exploring the prevalence rates of PTSD were conducted with subjects representing the general public as well as members of high-risk populations, for example Vietnam War veterans. American studies showed that PTSD is the fourth most frequent mental illness. However, only one in 20 patients seeks help for his or her difficulties (Berger, 2004).

The lifetime prevalence of exposure to trauma varies widely across traumatic events according to Breslau, Kessler, Chilchoat, Schultz, Davis, and Andreski (1998). Some occur in less than 2 percent (e.g. held captive, tortured, kidnapped, military combat) and others in more than 25 percent (e.g. being mugged, serious car accident). Sudden unexpected death of a close relative or friend is the most prevalent at 60 percent. The lifetime prevalence of exposure to any kind of trauma, according to DSM-IV, is 89.6 percent. The lifetime prevalence of PTSD varies between one and nine percent, depending on the kind of study being conducted (Steil & Ehlers, 1998). The differences in the prevalence rates are caused by methodical differences. For example, some studies were conducted prospectively and others retrospectively, some analyzed PTSD by diagnostic category and others on a continuous score, some assessed PTSD by interview and some by questionnaire, and some used DSM-III whereas others used DSM-III-R or DSM-IV criteria. Moreover, the frequency of traumatic events in different countries or regions varies, which was often not reported in older studies (Berger, 2004).

Moreover, according to Kolb (1989), many cases might go unrecognized by both the medical and psychiatric communities. PTSD is often confused with other anxiety disorders, alcoholism and various depressive reactions. Obviously, most PTSD sufferers wish to avoid all enquiries concerning their traumatic experiences and seldom spontaneously offer information.

2.5.1 Community-Based Studies

The two earliest community prevalence PTSD studies were conducted as part of the Epidemiological United States Catchment Area (ECA) study in St Louis (Helzer, Robins & McEvoy, 1987) and North Carolina (Davidson, Hughes, Blazer & George, 1991). Using DSM-III criteria, these studies found lifetime prevalence rates of PTSD in the total population of only 1.0 percent in St Louis and 1.3 percent in North Carolina. In St Louis, the prevalence rate for civilians exposed to physical attack was 3.5 percent and 20 percent for wounded Vietnam veterans. In North Carolina, approximately 6.6 percent of the population had sub-threshold PTSD. As another part of the ECA project, Robins, Fischbach, Smith, Cottler, Solomon, and Goldring (1986) reported that victims of dioxin, flooding, tornadoes, and radioactive well-water rarely exhibit new cases of PTSD. This lack of agreement with previous studies might lie in the mild nature of the disasters experienced or in the stringent criteria of DSM-III requiring numbing symptoms, which few victims show. It might also attest to the resiliency of the victims. Although not included in the ECA project, the community of Times Beach in the St. Louis area was of particular interest to Smith, Robins, Przybeck, Goldring, and Solomon (1986), as the flooding there was quite severe and was followed by the discovery of dioxin. The residents were evacuated and permanently relocated. Disaster victims showed significantly higher proportions with symptoms of six disorders. They exceeded both the unexposed and indirectly exposed groups in symptoms of depression, somatization, phobia, generalized anxiety, PTSD, and alcohol abuse. However, there is no indication of whether these were new symptoms or whether the symptoms had already existed before the disaster. When the onset of new symptoms was determined, only two significant differences were found: in depression and in PTSD

In contrast, one of the most recent studies, the replication of the "National Comorbidity Survey" conducted by Kessler, Berglund, Demler, Jin, and Walters (2005), resulted in a prevalence rate of 6.8 percent. In an earlier study with a representative sample of the general public with 5,877 participants aged 15 to 54 years, Kessler et al. (1995) estimated the lifetime prevalence rate to be 7.8 percent, using the DSM-III-R criteria. The authors also found that men experience trauma more frequently than women, but women develop PTSD twice as often as men (lifetime prevalence of 10.4 percent versus 5 percent). When a trauma has occurred, the probability of developing PTSD is 20.4 percent for women and 8.2 perccent for men in general. According to Ehlers (2000), women are more likely to be exposed to traumatic events than men, for example childhood sexual abuse or rape.

Surveys of the general population indicate that approximately 15 to 24 per cent of individuals exposed to traumatic events suffer from PTSD (Breslau, 2001a). In a study conducted by Breslau, Davis, Andreski, and Peterson (1991) on a random sample of 1,007 young adults (21-30 years of age) in southeast Michigan, the rate of DSM-III-R PTSD in individuals exposed to a traumatic event was found to be 23.6 percent, equivalent to a lifetime prevalence of 9.2 percent. For women, the prevalence rate was 11.3 percent compared to 6.0 percent for men. Despite the lower prevalence of exposure to trauma among women than men, of those women who were exposed, 30.7 percent had PTSD compared to 14 percent of exposed men. In the 1996 Detroit Area Survey of Trauma, Breslau et al. (1998) found the conditional risk of PTSD following exposure to trauma to be 9.2 percent. The highest risk of PTSD was associated with assaultive violence (20.9%) and the trauma most often reported as the precipitating event for PTSD (31% of all cases) was the sudden unexpected death of a loved one. This event was experienced by 60 percent of the sample and lead to a moderate risk of PTSD (14.3%). Women were at higher risk than men of developing PTSD, controlling for the type of trauma.

Other events causing traumatization can be accidents of all sorts, criminal assaults, or being admitted to a psychiatric hospital (Berger, 2004).

2.5.2 Studies with High-Risk Groups

An often cited study, the "National Vietnam Readjustment Study" showed a lifetime prevalence PTSD rate of 30 percent for male soldiers and 26 percent for female soldiers. The lifetime prevalence rate for partial PTSD, however, was 22 percent (Kulka, Schlenger, Fairbank, Hough, Jordan, Marmar & Weiss, 1990).

McFarlane observed a 27 to 32 percent prevalence of PTSD symptoms among Australian firefighters (McFarlane, 1989), using the General Health Questionnaire.

However, the decision to choose a position that might put oneself at risk also means that many individuals chosen for combat roles have a particular character type. In this way, the study of servicemen and highly selected occupational groups may lead to very atypical results that could minimize the role of vulnerability factors in PTSD (McFarlane, 1990).

2.5.3 Studies with Disaster Victims

Reported prevalence rates are obviously as diverse as the different samples in which they were solicited. For example, reports of PTSD after natural disasters have ranged from 1.5 to 67 percent (Wang, Gao, Shinfuku, Zhang, Zhao, & Shen, 2000). Due to the diversity of the samples, it is difficult to discern whether such differences are the result of methodological and sampling differences or studying a disorder with significant cultural variability (Yeomans & Forman, 2009).

One month after a mass shooting in a cafeteria in Texas, United States, 136 survivors were interviewed using the Diagnostic Interview Schedule/Disaster Supplement created by North, Smith, and Spitznagel (1994). Twenty percent of the men and 36 percent of the women met the DSM-III-R criteria for PTSD, which was the most prevalent psychiatric disorder. However, subsyndromal symptoms were abundant, and only 3.4 percent of the men and 1.5 percent of the women denied having any symptoms. Intrusive symptoms were the most frequent and numbing the least.

The prevalence rates also depend on the kind of trauma that has taken place. Norris (1992) examined the frequency and impact of ten potentially traumatic events in a sample of 1,000 adults. The primary purpose of this study was to investigate the long-term consequences of Hurricane Hugo, a natural disaster that had devastated large areas of North and South Carolina on September 22, 1989. Excluding Hurricane Hugo, sicty-nine percent of the participants experienced at least one of the ten events throughout their life. A tragic death occurred most often (30.2% of the participants) and sexual assault yielded the highest rates of PTSD (14% of women who were exposed to sexual assault developed PTSD). Gender differences in lifetime frequencies of events were also found. For example, women were more likely to be sexually assaulted, and men were more likely to be involved in motor vehicle accidents, to be physically assaulted, or to have experienced a violent event. In the total sample, 5.1 percent satisfied all four DSM-III-R criteria for current PTSD for reasons other than Hurricane Hugo. When cases that could be attributed specifically to Hugo were included, this number expanded to 6.2 percent. However, the proportions currently meeting the diagnostic criteria were primarily determined by criterion C. For example, 83 percent of Hurricane Hugo victims met criterion B (intrusion symptoms), and 42 percent met criterion D (arousal symptoms), but only 6 percent met criterion C (numbing symptoms). Figures for criterion C are low because three separate symptoms must be shown. After the hurricane, forty percent showed at least one symptom and 16 percent showed two symptoms.

Natural disasters usually lead to lower prevalence rates than man-made, intentional traumas. One of the first attempts to classify the psychological effects of disaster was made after the 1963 Skopje (former Yugoslavia) earthquake, when it was estimated that 75 percent of the survivors were suffering from mild disturbances and 10 percent manifested severe psychopathological reactions (Alexander, 1999). In 1987, Madakasira and O'Brien found that 59 percent of the interviewed survivors of a tornado qualified for a diagnosis of PTSD, according to DSM-III criteria, of whom 16 percent had a severe form. According to Kessler et al. (1995), the probability of developing PTSD after a natural disaster or fire is 5.4 percent for women and 3.7 percent for men (Kessler et al., 1995). Breslau et al. (1998) also measured the PTSD prevalence rate of victims of natural disasters and fires, and concluded that 0 percent of the women and 7.3 percent of the men suffered from PTSD after such incidences. However, according to Butollo and Hagl (2003), it is necessary to use caution when interpreting their results. In another study, Green, Lingy, Grace, and Leonard (1992) found that 20 years after a devastating flood, twenty-five percent of the affected people still suffered from PTSD.

In a study with victims of natural disasters, Steinglass and Gerrity (1990) investigated family and individual response to such disasters by assessing the psychosocial adjustment in adults who were relocated following a disaster in two different communities (tornado and flood victims). The psychosocial adjustment was measured at two points in time: at four months and at six months post-disaster. Horowitz's Impact of Event Scale (IES) and the Diagnostic Interview Schedule (DIS) for DSM-III were used. The results indicate that levels of short-term stress symptomatology and diagnosable PTSD were substantial in both communities (the four-month post-disaster findings indicate that, with few exceptions, at least half of the sample in each community reported experiencing each of the PTSD symptoms listed in the IES). At 16 months, the symptoms had decreased significantly. However, at 16 months, 4.5 percent of the tornado sample and 21.0 percent of the flood sample still met DSM-III criteria for PTSD.

North, Kawasaki, Spitznagel, and Hong (2004) assessed flood-related PTSD four and sixteen months after the Great Midwestern Floods of 1993 in the St. Louis area. Twenty-two percent of the flood survivors fulfilled the symptoms of PTSD at index and 16 percent at follow-up, measured with the Diagnostic Interview Schedule/Disaster Supplement (DIS-III-R). The most frequently endorsed PTSD symptoms at index were intrusive recollections (58%), insomnia (51%), difficulty concentrating (48%), and irritability (48%). Sixty-seven percent of the sample (N=162) met criterion B (intrusion), 28 percent met criterion C (numbing), and 56 percent met criterion D (arousal).

Russoniello, Skalko, O'Brien, McGhee, Bingham-Alexander, and Beatly (2002) assessed the level of PTSD experienced by fourth-grade children six months after Hurricane Floyd in North Carolina, USA. Ninety-five percent of the children experienced at least mild symptoms, and 71 percent had symptoms that were moderate to severe.

Bravo, Rubio-Stipec, Canino, Woodbury, and Ribera (1990) evaluated the psychological sequelae of the 1985 floods and mudslides in the Caribbean island of Puerto Rico. Since the event happened a year after an island-wide mental health survey was completed, the impact could be assessed prospectively as well as retrospectively. Multiple regression analyses with the total sample's retrospective data showed that, even after accounting for gender, age, education, pre-disaster symptom levels, and interview status, the higher the degree of disaster exposure, the higher the level of depressive, somatic, and PTSD symptoms. The sample's prospective data tend to confirm these results for depressive and somatic symptoms. The authors hereby used the emergence of new symptoms as a determinant for psychological dysfunction and found a significant, yet small disaster effect. It seems that individuals can be very resilient to the development of new psychiatric symptoms, even in the presence of severe stressors such as life threat, heavy personal and material losses, as well as evacuation and relocation. However, Canino, Bravo, Rubio-Stipec, and Woodbury (1990) assessed the impact of the disaster on the same sample regarding daily functioning and health, and found a significantly greater use of general health services for mental health problems among the exposed (21.5%) compared to the unexposed (12.4%). Those exposed to the disaster also reported poor health more frequently than the unexposed (7.5% versus 3.7%). Nevertheless, these findings need to be interpreted with caution, as the exposed and unexposed group might have differed in this regard before the disaster.

2.5.4 Studies in Non-Industrialized Settings

Prevalence studies of samples in non-industrialized countries are very scarce. In a comprehensive review, DeGirolamo and McFarlane (1996b) found that only six percent (8 out of 135) of prevalence studies were conducted with samples in those countries. For example, McCall and Resick (2003) investigated traumatic stress symptoms in a study with a sample of victims of domestic violence among the Ju/'hoansi (Kalahari Bushmen) of Eastern Namibia. This ethnic group is one of the last still transitioning from a hunter-gatherer lifestyle. In this sample, thirty-five percent met the criteria for PTSD and 85 percent reported at least some avoidance/numbing symptoms, but not to the extent that the DSM-IV criteria were fully met. Dyregrov et al. (2000) used the 22-item IES-R to assess the experiences and reactions of a large sample of Rwandan children one year after the Rwandan Genocide, and 79 percent exceeded the cut-off for PTSD (score of 17 or higher). These results can be compared to the study by Raymond (2000), in which 99 percent of internally displaced persons in Sierra Leone exceed the cut-off score on the IES. In this way, studies using quantitative methods conducted in African countries have reported varying PTSD prevalence rates, ranging from 35 to 99 percent.

In a review of 36 studies on disasters in non-industrialized regions, the findings suggest that natural disasters tend to be more destructive when they happen in non-industrialized settings, producing severe psychological consequences among the local community members (Norris, 2005).

A sample of displaced Bhutanese torture victims who were living in refugee camps in Nepal showed a much higher rate of PTSD than a highly similar cohort that had no torture history (14% versus 3%) and also significantly higher cumulative HSCL-25 anxiety and depression scores. Results of hierarchical logistic regression analyses showed that torture predicted symptoms of PTSD, depression, and anxiety (Shrestha, Sharma, van Ommeren, Regmi, Makaju, & Komproe, 1998). In a prevalence study of PTSD and depressive symptoms among the victims of the Kashmir conflict in Indian Kashmir, Yaswi and Haque (2008) found significantly more PTSD and depressive symptoms in participants directly affected by the conflict compared to participants indirectly affected. In addition, Tibetan refugee children living in Dharamsala, India, who had encountered severe traumatic events on their way to exile and also in their home country, showed a PTSD prevalence rate of 11.5 percent according to DSM-IV criteria. The prevalence of suspected PTSD in children who were reluctant to provide information about their feelings, but who gave the clinical impression of suffering from PTSD symptoms, was an additional 18 percent (Servan-Schreiber, Le Lin, & Birmaher, 1998). Wang et al. (2000) conducted a longitudinal study to assess the rates of earthquake-related PTSD in a community sample in northern China. The results suggest that the village with a higher level of earthquake exposure and higher level of post-earthquake support had lower PTSD levels than the village with a lower level of exposure but less post-earthquake support. The rate of onset of DSM-IV PTSD symptoms within nine months of the earthquake was 19.8 percent and 30.3 percent, respectively. The DSM-IV criterion requiring clinically significant distress or impairment in functioning for a diagnosis of PTSD was a major contributor to the lower rates of diagnosed PTSD. Otherwise, the pooled rate of PTSD in the two villages would have been 35.0 percent. According to DSM-III-R criteria, the pooled rate of PTSD would have been 41.4 percent.

One year after the super-cyclone in Orissa, India, PTSD according to the ICD-10 (WHO, 1992) criteria was diagnosed in 30.6 percent of a sample of children and adolescents affected by the cyclone. An additional 13.6 percent could be categorized as suffering from sub-syndromal PTSD. In higher exposure areas, more participants fulfilled the ICD-10 criteria for PTSD than in lower exposure areas (43.7% versus 11.2%) (Kar, Mohapatra, Nayak, Pattanaik, Swain, & Kar, 2007).

After the 2004 tsunami, destruction along the coastlines of southern India was massive, and according to Carballo, Heal, and Horbaty (2006), depression was a common reaction reported by the survivors. Although high rates of PTSD were expected, it is not clear how common the disorder really was. Few cases of PTSD were diagnosed in India in the weeks following the disaster, which was attributed to people's natural resilience and ability to cope when faced with a disaster (Carballo et al., 2006). A study by Kumar, et al. (2007) with 314 victims of the 2004 Asian tsunami in Tamil Nadu revealed a PTSD prevalence rate of 12.7 percent, measured using the Harvard Trauma Questionnaire two months after the tsunami. Intrusive thoughts and sleep disturbances were the most, and irritability and emotional numbness the least frequently reported symptoms. Among individuals who sought help from psychiatrists (15% of the overall sample), 31.9 percent were diagnosed with PTSD and 48.9 percent with major depression. In a recent study with Sri Lankan tsunami victims by Hollifield, Hewage, Gunawardena, Kodituwakku, Bopagoda, and Weerarathnege (2008), the prevalence of clinically significant PTSD, depression, and anxiety was 21 percent, 16 percent, and 30 percent, respectively. Measures used were an adapted questionnaire assessing PTSD according to ICD-10 and the Hopkins Symptom Checklist-25 (HSCL-25). Moreover, participants reported a mean of eight somatic complaints associated with psychological distress.

However, Kokai, Fujii, Shinfuku, & Edwards (2004) found relatively low prevalence rates of post-traumatic stress disorder (PTSD) in Asian countries compared to other countries, such as Mexico or Azerbaijan. They believe the reason for this phenomenon is strong family and community ties, as well as the lower expression of psychological content, and inclination to somatize distress.

Kagee (2005) investigated the prevalence of psychological distress and PTSD in a sample of South African political prisoners who were subjected to torture and abuse. Post-traumatic stress was assessed with the Impact of Event Scale (Horowitz, et al., 1979) and 17.57 percent scored above the clinical cut-off point, which the author set at 44. According to the author's interpretation, a score of 0-8 is in the subclinical range, 9-25 in the mild range, 26-43 in the moderate range, and 44 and higher in the severe range. The mean score of the IES was 36.69, which therefore falls in the moderate range of PTSD symptoms.

In other countries, some prevalence studies examined possible cultural differences. For example, Howard, Loberiza, Pfohl, Thorne, Magpanty, and Woolson (1999) found no significant differences in the prevalence rates of PTSD or other psychiatric disorders between tribal and non-tribal disaster victims six years after the eruption of Mt. Pinatubo in the Philippines. PTSD was the most prevalent disorder between the two groups with a prevalence rate of 27.6 percent (32.2% in the tribal group). Mental illness prevalence rates in both Filipino ethnic groups were comparable to those found in an American study using the same assessment instrument (PTSD Checklist).

In a study by Witruk, Reschke, and Stück (2009), psychological consequences and the effects of first aid psychology on tsunami victims were investigated in Sri Lanka and Indonesia. People affected by the tsunami and people not affected participated in the study. It was shown that both groups suffered mild to severe PTSD symptoms, as was measured with the IES by Horowitz et al. (1979). However, the extent of traumatization was higher in the tsunami-affected group (mean score of 34.00 compared to 18.27). A total of 20.8 percent of the tsunami victims suffered from severe PTSD, whereas only 5.4 percent of the non-affected group suffered from severe PTSD.

2.6 Prognosis and Duration of PTSD

The probability of developing PTSD depends on the kind of trauma that happened. After a motor vehicle accident, for example, the probability of developing PTSD is 1 to 49 percent, whereas after a rape, it is 50 to 90 percent. Often the PTSD symptoms disappear after a few weeks, but if they continue for more than three months, the prognosis is unfavorable, as there is an increasing danger of the symptoms becoming chronic (Frommenberger & Maercker, 2006). Breslau and Davis (1992) performed an analysis of data gathered from 394 respondents (of a sample of 1007 young adults) who reported traumatic events. Ninety-three of the respondents met the criteria for PTSD, and of these, 57 percent had persistent symptoms for more than a year. Moreover, they investigated factors that are possibly associated with an increased risk of the PTSD symptoms becoming chronic. The individuals with chronic PTSD had a higher total number of symptoms and showed higher rates of numbing and hyperreactivity to stressor stimuli, anxiety or affective disorders, and other coexisting medical conditions. A family history of antisocial behavior and female gender were found to be factors specifically related to chronic PTSD (Breslau & Davis, 1992). According to Breslau (2001a), eighty-two percent of the people suffering from PTSD meet the DSM-IV criteria for chronicity, meaning that their symptoms last more than three months. Approximately 74 percent continue to have symptoms for six months or more (Breslau, 2001a). The symptoms become chronic in a quarter of the cases according to Hidalgo and Davidson (2000).

Moreover, in the National Comorbidity Survey by Kessler and colleagues (1995), survival curves based on the duration of symptoms decreased most steeply in the first 12 months after the onset of symptoms. The curves continued to decline with a more gradual slope for approximately six years after symptom onset. The median time to remission was 36 months among the respondents who sought professional treatment and 64 months among those who did not. PTSD failed to remit in more than one-third of the respondents even after many years, both in the treatment subsample and in the subsample of respondents who never received treatment. According to Breslau et al. (1998), in more than one-third of the cases the symptoms persisted for more than five years.

The retrospective study conducted by Breslau et al. (1998) showed a median time to remission of PTSD of 24.9 months. However, the median duration appears to differ by gender. The authors reported a median duration of PTSD in women approximately four times as long as that in men (48.1 months versus 12.0 months) and a hazard ratio for remission in women versus men of 0.53. For both men and women, the symptoms persist longer if the traumatic event is experienced personally, as opposed to traumatic events that others have experienced. Women suffer from PTSD symptoms for 14 months if the event is experienced by others (e.g. the death of a loved one) and for as long as six years if they experience the trauma personally, while the figures are six months versus two years for men (Breslau et al., 1998).

The prevalence and longitudinal course of PTSD after a natural disaster were studied by McFarlane (1986), who interviewed a group of 459 firefighters exposed to the Ash Wednesday bushfires in South Australia. The level of morbidity four months after the fire (21%) remained almost unchanged 29 months after the disaster.

In a prospective longitudinal study with 967 patients who attended an emergency clinic after a motor vehicle accident, at three months, 23.1 percent of the sample met DSM-IV criteria for PTSD and at one year, 16.5 percent met the criteria. Approximately half of the participants met the re-experiencing, hyperarousal, and disability criteria at both time points, but whether the diagnosis was given depended on the avoidance and numbing criterion. Participants meeting the diagnostic criteria for PTSD at three months had a 50.3 percent chance of still suffering from PTSD at one year. The most important factors related to chronicity were objective measures of trauma severity, perceived threat, dissociation during the accident, female gender, previous emotional problems, and litigation. Maintaining psychological factors, i.e. negative interpretations of intrusions, rumination, thought suppression, and anger cognitions, enhanced the accuracy of the prediction. Rumination, anger cognition, injury severity, and prior emotional problems identified cases of delayed onset (Ehlers, Mayou & Bryant, 1998).

2.7 Comorbidity with other Mental Disorders

The physical and psychological reactions to a trauma are quite diverse. PTSD is only one of many different possible psychological reactions to a traumatic situation. Other mental disorders that can occur after a trauma are major depression, panic disorder, generalized anxiety disorder, phobic avoidance behavior, or substance abuse. Physical symptoms can be hypertonus, asthma, or chronic pain syndrome (Yehuda, 2002).

Various studies demonstrated that PTSD often accompanies other disorders such as anxiety and affective disorders, substance abuse/dependence, somatization, as well as antisocial personality disorder (Breslau, Davis, Petersen & Schulz, 1997). It is not surprising, as a high degree of overlap exists between the symptoms of PTSD and depression. For example, PTSD shares 10 of the 17 symptoms that constitute the Hamilton Rating Scale for Depression. Nevertheless, these similarities should not obscure the fact that PTSD, depression, and anxiety all show sufficient differences to be considered distinct conditions (Yule, 2001). In an earlier study, Breslau et al. (1991) found that the two most prevalent disorders among persons suffering from PTSD were major depression and alcohol abuse/dependence, diagnosed in respectively 36.6 percent and 31.2 percent of the PTSD group. A total of 82.8 percent of persons with PTSD had one or more other psychiatric disorders. These figures resemble those of the National Comorbidity Survey, where it was found that 79 percent of the women and 88.3 percent of the men suffering from PTSD had at least one other mental disorder throughout their life (Kessler et al., 1995). Breslau et al. (1997) later conducted a study in a sample of 801 mothers (mean age = 33.1 years). In the total sample, the lifetime prevalence of PTSD was 13.8 percent. Those subjects with PTSD had significantly higher rates of major depression, alcohol abuse/dependence, and drug abuse/dependence than those without PTSD. Major depression was the most prevalent comorbid disorder, occurring in 43.2 percent of the women with PTSD. Altogether, the lifetime comorbidity of PTSD with the aforementioned disorders was found to be 73 percent. Another goal of the study was to determine whether preexisting psychiatric disorders could lead to an increased risk of PTSD by increasing both the likelihood of trauma exposure and the subsequent vulnerability to the disorder. These findings have been confirmed, as the most plausible hypothesis for drug abuse following PTSD is the individual's effort to reduce the symptoms by self-medication (Breslau, 2001a).

Gerrity and Steinglass (1994) interviewed 40 families who had lost their homes due to a severe flood in West Virginia to assess relocation stress. The results of the study suggest short- and long-term psychopathology in almost all cases among both adults and children. The conductors of the study reported short-term reactions in adults, including psychic numbing, sluggishness in thinking and decision-making, anxiety, grief, despair, and severe sleep disturbances. Long-term responses (two years post-disaster) included physical complaints, survivor guilt, listlessness, apathy, decreased social interaction, and chronic depression.

According to Alexander (1999), depressive symptoms will be highest among those victims of natural disasters who are bereaved, severely injured, suffering from mental illness before the disaster, weak socio-economically and whose sources of social support are very restricted. Shock, numbness, anxiety, guilt, somatic complaint (e.g. loss of appetite or sleep), and even physical deterioration and increased mortality may accompany depression.

Green et al. (1992) interviewed 193 survivors of the breach in the Buffalo-Creek dyke. Fourteen years after the flood, 25 percent of the population still showed symptoms of PTSD. Forty-two percent of these PTSD patients developed comorbidity with depression, 42 percent with generalized anxiety disorder, and 29 percent with simple phobia. The "National Vietnam Veterans Readjustment Study" showed that 20 years after the Vietnam War, fifteen percent of the veterans still suffered from PTSD symptoms. Of these 15 percent, seventy-three percent developed substance abuse, 26 percent major depression and 21 percent dysthymia. Thirty-one percent developed antisocial personality disorder, and 98.8 percent of Vietnam veterans with PTSD had a history of some other DSM-III-R disorder, compared with 40.6 percent of those without PTSD. The study also showed that those veterans who experienced extremely traumatic events at the war front complain more about chronic health problems and assess their overall health as worse than those veterans who were not sent to the war front (Kulka et al., 1990). In a sample of 773 motor vehicle accident victims, twenty-three percent of the affected people showed symptoms of PTSD three months after the accident, 22 percent developed a traveling phobia, 19 percent suffered from generalized anxiety and 5 percent suffered from depression (Mayou, Bryant & Ehlers, 2001).

In addition to a high risk of comorbidity, an increased risk of numerous physical complaints can also occur. In a study by Andreski, Chilcoat and Breslau (1998), individuals with PTSD experienced more than three times as many physical symptoms associated with somatization disorder than those without PTSD (24.7% versus 8.2%).

North et al. (2004) found that the most prevalent post-disaster diagnosis identified four months after the Great Midwestern Floods of 1993 was flood-related PTSD (22%), followed by major depression (20.4%). More than half of the individuals with flood-related PTSD (60%) had a comorbid disorder after the floods, although only 20 percent of people without PTSD had another post-disaster diagnosis. One half (51%) of those with and 12 percent of those without flood-related PTSD were diagnosed with major depression after the flood. Panic disorder was present in 9 percent of those with PTSD and 1 percent of others, and generalized anxiety disorder was present in 12 percent of those with PTSD and 2 percent of others. Furthermore, new somatoform symptoms were reported by 53 percent of those with PTSD and 17 percent of others. In a follow-up study conducted 16 months after the flood, 15.8 percent of the participants had a current diagnosis. Less than one third (29%) of those diagnosed with PTSD at index or follow-up had fully recovered from it. The recovery rate for major depression was 62 percent. Of those with major depression at either index or follow-up, who also had comorbid PTSD, only 13 percent recovered, compared with 86 percent of the individuals with PTSD but without major depression.

When investigating the effects of a mass shooting, North, Smith, and Spitznagel (1994) found that 50 percent of the women and 25 percent of the men with post-disaster PTSD had comorbidity with at least one other disorder, especially major depression (35% of the women and 25% of the men with post-disaster PTSD). A total of 72.7 percent of the women and 100 percent of the men with major depression after the disaster met the criteria for PTSD. Only a few of the participants who did not show any PTSD symptoms developed any other disorder (9% of the women and 7% of the men).

PTSD and depression are among the most serious problems for populations exposed to natural disasters. Armenian, Morikawa, Melkonian, Hovanesian, Akiskal, and Akiskal (2002) assessed the rate of major depression in adult survivors of the 1988 earthquake in Armenia. A total of 1,785 participants were interviewed approximately two years after the earthquake, using a special questionnaire based on the National Institute of Mental Health (NIMH) Disaster Interview Schedule/Disaster Supplement. In total, 52 percent of the sample fulfilled the DSM-III-R criteria for major depression. Most of the earthquake survivors fulfilling the criteria of depression also suffered from comorbidity. In particular, 73.4 percent of these persons also showed the PTSD symptoms according to the DSM-III-R criteria. Cases of having only depression as a diagnosis were compared with cases without any psychiatric diagnosis. The first group contained more females and residents of the worst affected area. The risk for depression increased with the amount of loss. In a previous publication on PTSD in Armenia, the authors found that the intensity of loss was a determinant of PTSD, and that early social support provided to survivors reduces the risk of PTSD (Armenian, Morikava, Melkonian, Hovanesian, Haroutunien, Saigh, et al., 2000).

Norris, Friedman, Watson, Byrne, Diaz, & Kaniasty (2002) conducted an empirical review of the empirical research that has been published over the last twenty years. Results for 160 samples of more than 60,000 disaster victims were coded by sample type, disaster type, disaster location, outcomes and risk factors observed, and the overall severity of impairment. Outcomes included specific psychological problems, non-specific distress, health problems, and chronic problems in living, resource loss, and problems specific to youth. Regression analyses showed that disaster victims were more likely to show impairment if the sample was composed of youth rather than adults, was from developing rather than developed countries, or experienced intentional, man-made violence (e.g. terrorism, mass-shootings) rather than natural or technological disasters. Interestingly, most samples of rescue workers showed remarkable resiliency. Within adult samples, more severe exposure, female gender, middle age, ethnic minority status, secondary stressors, prior psychiatric problems, and weak or deteriorating psychosocial resources consistently increased the likelihood for adverse outcomes. The most commonly observed psychiatric disorder was PTSD, as 68 percent of the samples investigated the levels of PTSD. The second most commonly observed psychiatric problem was depression, identified in 58 percent of the samples. Anxiety was identified in 20 percent of the samples and non-specific distress, measured with the Brief Symptom Inventory, was identified in 39 percent of the samples. Another set of outcomes dealt with health problems and concerns, which were identified in 36 percent of the samples. Moreover, chronic problems in living have rarely been assessed, but where they have been assessed, they have typically been observed (10%). In the months following a disaster, disaster victims are more likely than non-victims to experience hassles or life events that serve as stressors in their own right. These stressors act as mediators between acute exposure and chronic psychological effects. Psychosocial resource loss was explicitly identified in 9 percent. Although some studies have used global indices of resource loss, others have observed a decline in specific resources, such as perceived social support, social embeddedness, self-efficacy, optimism, and perceived control. Social resources are vulnerable to the impact of disaster, as disasters remove significant supporters from victims' networks through death. Temporary or permanent relocation disrupts neighborhood patterns and engenders interpersonal strains and conflicts. In addition, support providers are victims themselves after a disaster.

3 Risk Factors and Predictors of PTSD

The question arises, "What turns a stressful experience into a traumatic one?" Different people deal with similar experiences in different ways, and only a small percentage of people suffer from enduring psychological distress. Whether someone develops PTSD or not depends on internal and external processes, which are influenced by aspects of the traumatic experience, as well as personality and situational factors (Krüsmann & Müller-Cyran, 2005).

Given that approximately 50 percent to 60 percent of the U.S. population is exposed to traumatic stress but only 5 to 10 percent develop PTSD, which systematic risk factors identify who will or will not develop PTSD (Ozer, et al., 2003)?

Various factors can increase the risk of developing PTSD after experiencing a trauma. Examples are the subjective experienced loss of control or the expectation of adverse consequences (Maercker, Schützwohl & Beauducel, 2000). Not only cognitive, but also demographic and personality factors increase the probability of suffering from PTSD, such as low intelligence or a lack of education, being female, a pre-existing psychiatric illness or being young (meta-analysis by Brewin, Andrew & Valentine, 2000), or old age (Maercker, 1998; Phifer & Norris, 1989). Peri- and post-traumatic factors (e.g. severity of the trauma, social support, and additional stress of everyday life) play a greater role than pre-traumatic factors (e.g. education, age, gender, pre-existing mental illness) (Brewin et al., 2000; Maercker, 2003). Loss of resources (such as feeling of control over one's life, a sense of optimism, or feeling independent) is central in predicting psychological distress after a natural disaster, according to Freedy, Saladin, Kilpatrick, Resnick, and Saunders (1994). Certain coping strategies such as emotion-centered coping and distancing behavior have also been found to be predictors of PTSD (Solomon, et al., 1988). In addition, a lack of social support has been proven to be a serious risk factor (Brewin et al., 2000; Breslau et al., 1998).

3.1 Socio-Demographic Factors

When studying an urban population of young adults in southeast Michigan, Breslau et al. (1991) found that 39.1 percent of the respondents reported exposure to one or more traumatic events. Exposure was more common in men (43.0%) than in women (36.7%), but was unrelated to race or marital status. Other risk factors for exposure to traumatic events were low education (i.e. less than a college education), a history of childhood conduct problems, neuroticism, extraversion, and a family history of psychiatric disorders or substance problems. Risk factors for developing PTSD after exposure to traumatic events included female gender, early separation from parents, neuroticism, pre-existing anxiety or depression, and a family history of anxiety.

Gender comparisons across the different types of traumatic events showed that assault and sudden injury or accident affected men and women equally. However, seeing someone killed or seriously injured and news of the sudden death of a close relative or friend led to significant higher PTSD rates in exposed women than men. Therefore, with the exception of rape, women's greater vulnerability to traumatic events was for the most part confined to violent events happening to others (Breslau et al., 1991).

Trauma and subsequent risk for PTSD (as defined in the DSM-IV) was additionally examined in a community sample of 2,181 respondents aged 18 to 45 years in the Detroit area by Breslau et al. (1998). Overall estimates of the lifetime prevalence of PTSD in women have been found to be approximately twice those in men (13% versus 6.2%). In studies conducted before the introduction of the DSM-IV criteria, estimates ranged from 10 to 14 percent. The Detroit Area Survey reports a lifetime prevalence of PTSD in women of 13 percent. Moreover, the mean duration of the PTSD symptoms is almost four times greater in women (mean = 48 months) than in men (mean = 12 months). These findings may be due to the fact that females are at greater risk of PTSD than males when exposed to assaultive violence, and this gender difference persists virtually unchanged when the females' higher exposure to rape is accounted for. Women respond to assaultive violence more with numbing and avoidance than men. Therefore the likelihood of developing PTSD according to the DSM-IV criteria is greater.

The findings of the National Comorbidity Survey suggest that the traumas most commonly associated with PTSD are combat exposure and witnessing among men and rape and sexual molestation among women. The authors imply that gender differences, i.e. the higher vulnerability of women to PTSD, may result from the fact that the most upsetting traumas nominated by women were associated with a higher probability of PTSD than those nominated by men. Moreover, lifetime PTSD was significantly more prevalent among the previously married (separated, divorced, or widowed) than the currently married for both men and women, controlling for age. Among men, but not women, PTSD was significantly more prevalent among currently married than never married (Kessler et al., 1995).

In the study conducted by Norris (1992), black men appeared to be most vulnerable to the effects of events when the impact was analyzed as a continuous variable (perceived stress), but young people showed the highest rates of PTSD. Women generally perceived their life as more stressful, and women who were victims of violent crime were more likely to satisfy diagnostic criteria for PTSD (12% compared to 6% men). In a later study by North et al. (2004), it was found that flood-related PTSD developed more than three times as often in people with (35%) than without (11%) a pre-existing psychiatric disorder. However, no demographic characteristics (gender, age, education, marital status) predicted PTSD.

In an exploratory study, the stress experiences of 20 flood victims in the Bloemfontein area of South Africa were investigated (Burger, van Staden, & Niuewoudt, 1989). Measures consisted of demographic variables, Rotter's Locus of Control Scale, Horowitz's Impact of Event Scale, and unstructured interviews. Female subjects reported higher stress levels than male subjects, although no differences in the subjects' experiences of stress were found with regard to age, occupation or the measure of internal-external locus of control.

Five months after a tornado devastated a rural community in North Carolina, Madakasira and O'Brien (1987) surveyed the mental health status of 116 disaster victims. Although a lack of social support was more often noted in victims with severe PTSD, other demographic factors such as age, gender, race, marital status, employment, education level, degree of injury, degree of property damage, or history of mental illness did not appear to correlate to the presence of PTSD.

A study by Koopman, Classen, and Spiegel (1994) investigated the relationships between dissociative symptoms experienced in the immediate aftermath of the 1991 Oakland/Berkeley firestorm and exposure to the disaster as well as appropriateness of coping strategies, stressful life events in the past, and demographic variables. A stepwise multiple regression analysis showed that respondents' dissociative and anxiety symptoms were significantly related to contact with the fire, having more stressful life events, and being female, but not to marital status, education, or age. People with less dissociative experience are more likely to engage in active coping strategies, whereas those with high dissociative experience engage more in passive and avoidant coping strategies. Moreover, people with greater dissociative experience are more likely to engage in inappropriate coping strategies such as trying to get closer to the fire or entering blocked-off areas.

3.1.1 Gender

One barrier to examining the issue of gender differences in clinical or "at-risk" populations is that a number of investigators focus on survivors of one gender, for example Vietnam War veterans, usually men, or rape or incest victims, usually women (Green, 1993).

Generally, female gender has proven to be a risk factor for developing PTSD. Prevalence rates are higher for women, often twice as high as for men, even though men are exposed to more traumatic events. Butollo and Hagl (2003) argue that men encounter more traumatic events, but the events that women experience are more severe. However, the gender effects still exist after controlling for the type of traumatic event in the studies by Breslau et al. (1998) and the meta-analysis by Brewin et al. (2000). A thorough analysis of the data from the Detroit Study (Breslau et al., 1998) revealed that the gender differences are due to the category of violent assaults (excluding rape). The authors postulate that women perceive violent assaults as more threatening. In Ehlers et al.'s study (1998) with motor vehicle accident victims, women had a greater risk of developing PTSD at three months (28.9% versus 18.2%), but there was no significant gender effect at the one-year follow-up.

These gender differences might also be due to the fact that women report more negative responses from family and friends after a traumatic event. Such negative support could mediate the relationship between gender and distress. Overall negative response and support satisfaction, but not positive support, were significantly associated with PTSD symptoms in a study with victims of violent crime in the United Kingdom. The effects of support satisfaction and negative response on symptoms were significantly greater for the female participants (Andrew, Brewin, & Rose, 2003).

However, the findings of a meta-analysis of Horowitz's Impact of Event Scale ratings presented during a 20-year period by different groups of trauma victims suggested that neither the trauma victim's gender nor the victim's age is critical for the manifestation of stress reactions after the event. In a series of hierarchical regression analysis, only the type of event and time elapsed since the event significantly predicted intrusion and avoidance symptoms, but not the variables of age and gender (Sundin et al., 2003). Furthermore, in Bravo et al.'s (1990) study on flood victims in Puerto Rico, differences in the various socio-demographic groups were discovered, and thus no gender, age, or education effects were found.

In a study conducted in the same geographical area as the area in which this dissertation study was conducted– Azheekkal, Kerala - by Sunidharan (2005), certain demographic factors and coping strategies were investigated to determine the risk factors for developing anxiety and depression. The target sample consisted of 62 survivors of the tsunami who were living in relief camps at the time of the study (one to three months after the tsunami). Depression was assessed with the Beck Depression Inventory (BDI) (Beck, Ward, Mendelsohn, Mock, & Erbaugh, 1961), anxiety with the Hamilton Anxiety Rating Scale (HARS) (Hamilton, 1959), and coping strategies with the Coping Checklist

(CCL) (Rao, Prabhu, & Subhakrishnan, 1989). Additionally, a demographic sheet was given to collect information about demographic- and trauma-related variables. The mean scores on the BDI and HARS of the participants whose homes were completely destroyed did not significantly differ from those whose homes were not completely destroyed. Widowed participants, however, showed a significantly higher score on the BDI as well as the HARS. There was a significant gender difference on both the anxiety and depression scores. This difference may be attributed to the gender roles in society. In Kerala, gender roles are often rigid and closely aligned with traditional gender stereotypes; men are expected to be strong, unemotional, and aggressive, whereas women are expected to be nurturing, emotional, and submissive. The gender difference might be a result of men and women expressing psychological distress in different ways. Correlation analysis showed that the higher the number of family members lost, the higher the depression score. A multiple regression was performed to find the predictor variables for anxiety and depression: the coping strategy "distraction" and the demographic variables female gender, widowed marital status, and substance abuse predict anxiety; whereas the coping strategy "faith" and the demographic variables female gender and number of family members lost predict depression.

3.1.2 Age

Psychological reactions to disaster are specific to the age of the victim. However, different studies have found different age groups to be at risk for PTSD: Young and middle-aged adults are at risk according to Norris (1992), and young and old age according to Maercker (1997, 1998). If children are unable to understand and rationalize the event, they may suffer from phobias, sleep disturbances, loss of interest in schoolwork and aggressive or undisciplined behavior. The elderly who live alone and lack adequate support systems may suffer from depression and a sense of hopelessness (Alexander, 1999). Moreover, older individuals might have more previous traumatic experience, which presents a risk factor in itself (Green, Wilson & Lindy, 1985; Breslau et al., 1998).

In contrast, according to Knight, Gatz, Heller, and Bengston (2000), older adults are more resilient to post-disaster distress than younger adults. They postulated that this might be due to older adults being less emotionally reactive to stressful events, including disasters (maturation hypothesis). Older adults might also be inoculated against strong emotional reaction through prior experience with disaster. In a longitudinal study conducted by Phifer and Norris (1989), the effect of exposure to the 1981 flood in southeastern Kentucky on older adults' response to the 1984 flood was analyzed, showing that those participants with prior experience showed greater psychological adaptation. However, the largest predictor of post-flood depression was pre-flood depression and older adults were lower on depression both before and after the flooding. Knight et al. (2000) collected longitudinal data from victims of the 1994 Northridge earthquake in southern California. The sample was categorized into three groups: middle-aged adults (30 to 54 years), young-old adults (55 to 75 years), and old-old adults (over 76 years). The young-old were less depressed both before and after the earthquake. The old-old group reported lower levels of earthquake-specific rumination with exposure level controlled. The extent of exposure to earthquake damage affected specific rumination about the earthquake, but not the level of depressive symptoms. Although prior experience with earthquakes was associated with lower depressive symptoms, it did not affect earthquake-specific rumination. Virtually no support was provided for the maturation hypothesis, and modest support was found for the inoculation hypothesis, with respect to depression.

However, in a study with 1,289 survivors of the floods in the North Indian state of Bihar on August 18, 2008, persons over the age of 60 had higher scores on PTSD, as well as depression (Telles, Singh & Joshi, 2009). Participants were screened with the Screening Questionnaire for Disaster Mental Health (SQD), which was used in Japan after the 1995 Great Hanshin earthquake and was shown to have good screening validity, especially for PTSD (Fujii, Kato & Maeda, 1995). Separate two-factor ANOVAs were used to compare persons of both genders and five different age groups. No significant differences between men and women were detected, but the oldest age group showed significantly higher scores compared to all of the other groups. In this way, the vulnerability of older persons was demonstrated (Telles et al., 2009).

3.1.3 Education

In the study by Armenian et al. (2000) with survivors of the 1988 earthquake in Armenia, the level of education was found to be inversely related to the risk of PTSD. This finding might be interpreted in such a way that some learned components of behavior interpretation might be used by the individual as a defense against the development of PTSD. Having less than a college education was also found to be a risk factor for developing PTSD, according to Breslau, et al. (1991).

3.1.4 Earlier psychiatric disorders

The study conducted by Breslau et al. (1997) confirmed that pre-existing psychiatric disorders could increase the risk of PTSD. Women who had preexisting major depression, any anxiety, or substance abuse/dependence were at increased risk of first exposure to traumatic events. In addition, the risk of developing PTSD symptoms was increased in exposed women who had prior major depression, independent of any relationship with other disorders. A connection between childhood trauma and the onset of PTSD symptoms after another trauma in adulthood was discovered by Nishith, Mechanic, and Resick (2000).

However, in the study by North et al. (1994), most participants meeting the criteria for PTSD had no prior history of psychiatric illness (59% of women and 83% of men). Pre-disaster rates of PTSD were quite high in this sample and did not predict PTSD after the disaster. Pre-disaster psychiatric disorders (excluding PTSD) only predicted higher rates of PTSD in women, but not in men. Forty-one percent of women with post-disaster PTSD had a pre-disaster psychiatric diagnosis (excluding PTSD), compared to 16 percent of those without PTSD.

3.1.5 Exposure

Hardin, Weinrich, Weinrich, Hardin, and Garrison (1994) investigated the effects of a natural disaster on adolescents. A total of 1,482 high school students who were exposed to Hurricane Hugo participated in the study. As exposure increased, so did symptoms of psychological distress, i.e. anger, depression, anxiety, and global mental distress. Female students also reported higher levels of distress than male students. Black students, who were most exposed to the disaster and to negative life events, reported the least psychological distress. In most cases, other stressful life events were at least as strong a predictor of psychological distress as was exposure to the hurricane. Self-efficacy and social support were found to be protective factors and associated with lower psychological distress, with the standardized beta coefficients being larger for social support than for self-efficacy.

In Russoniello et al.'s (2002) study with children affected by Hurricane Floyd, children whose homes were flooded were three times as likely to develop symptoms of PTSD as children whose homes were not flooded. One-hundred percent of the children with flooded homes reported symptoms of re experiencing the event, eighty-five percent reported symptoms of hyperarousal, and 64 percent reported symptoms of numbing/avoidance. Girls were twice as likely as boys to develop PTSD. In the study by Armenian et al. (2000), PTSD cases included more individuals from areas with the worst destruction by the earthquake.

3.2 Predictors of PTSD: A Meta-Analysis

Ozer et al. (2003) conducted a meta-analysis including 68 studies regarding seven factors predicting PTSD (prior trauma, prior psychological adjustment, family history of psychopathology, perceived life threat during the trauma, post-trauma social support, peritraumatic emotional responses, and peritraumatic dissociation. All predictors yielded significant effect sizes, with family history, prior trauma, and prior adjustment the smallest (weighted $r = .17$) and peritraumatic dissociation the largest (weighted $r = .35$). These results suggest that peritraumatic psychological processes as opposed to prior characteristics are the strongest predictors of PTSD.

An initial review of predictive factors for PTSD was published by Brewin et al. (2000). These authors conducted meta-analyses with 77 studies on 14 separate risk factors for PTSD: five demographic factors (age, gender, socioeconomic status, education, and race) and nine other variables. According to Ozer et al. (2003), these nine can be categorized into three groups: (1) historical or static person characteristics, (2) trauma severity, and (3) social support and intercurrent life stress. Factors such as psychiatric history, reported childhood abuse, and family psychiatric history had the most uniform predictive effects. Individually, the effect size of all the risk factors was modest, but factors operating during or after the traumatic event, such as severity of the trauma, lack of social support, and additional life stress had stronger effects than the pretrauma factors (Brewin et al., 2000). The strongest predictor of PTSD was found to be a lack of social support, with the weighted effect size being $r = .40$. In the 2003 meta-analysis conducted by Ozer et al., the weighted average correlation for 11 studies of the relationship between perceived social support following the trauma and PTSD symptoms was -.28, a significant effect size in the small-to-medium range. It is interesting to note that this negative relationship between social support and PTSD was strongest in studies in which more than three years had elapsed between the traumatic event and the assessment of PTSD (weighted $r = -.42$). Therefore, the findings suggest that the effects of social support are cumulative over time. All in all, stronger predictive factors yielding coefficients greater than .20 were more proximal to the traumatic event such as peritraumatic dissociation or perceived social support as opposed to more distal pre-traumatic factors such as characteristics of the individual or his or her life history (Ozer et al., 2003).

4 Protective Factors

Protective factors can be considered reversed risk factors. In this way, abundant material and social resources, above-average intelligence or self-enhancing, positive experiences during childhood can reduce the risk of developing PTSD (Butollo & Hagl, 2003). Moreover, a permanent positive relationship with at least one primary caregiver, growing up in an extended family, high intelligence, a robust, sociable and active temperament, secure attachment behavior, social support, reliable and supportive people to relate to in adulthood, and other constant social relationships are seen as protective factors throughout the life span, according to Egle, Hoffmann and Joraschky (1996).

4.1 Resilience, Hardiness, and Sense of Coherence in the Salutogenesis Model

In the 1970s, the main focal point of interest shifted from pathology to competence or resilience in the face of adverse situations. The correlation between stressful events and illness is typically only .30. Therefore, the emphasis on "stress and illness" research has begun to shift toward the study of "resistance and resources" that can mitigate the effects of stressors (Antonovsky, 1979). Such resources might include constitutional strengths (e.g. little history of family illness), social support (e.g. social contacts and status), health practices (e.g. jogging), and personality dispositions (Kobasa, Maddi & Kahn, 1982).

In the empirical literature, three main factors are often mentioned as efficient coping mechanisms for dealing with traumatic stressors: *resilience, hardiness* and *sense of coherence.* Resilience is a combination of the aforementioned factors and an ability to deal with stressors in a flexible way (for PTSD: Flach, 1990); hardiness is a mental construct consisting of (a) cognitive beliefs concerning the meaning of life, (b) cognitive beliefs concerning one's own influence on events, and (c) a positive attitude toward life's challenges (Kobasa, 1979); and sense of coherence is a result of cognitive beliefs about being able to cope with life and its challenges, and viewing life as meaningful rather than disconnected and arbitrary (Antonovsky, 1979).

4.1.1 Resilience

When the concept of stress was first devised, scientists hypothesized that an organism strives to maintain its own equilibrium or homeostasis. A healthy person automatically adjusts to external and internal variations to uphold this homeostasis (Flach, 1990). The first researcher to define the so-called stress syndrome was Hans Seyle (1988), who postulated that "Stress is the stereotyped part of the body's response to any demand... It is associated with the wear and tear on the human machinery that accompanies any vital activity". Seyle's "general adaptation syndrome" evolves in three stages: (1) the "alarm reaction", during which defensive forces are mobilized; (2) the "stage of resistance", which reflects full adaptation to the stressor; and (3) the "stage of exhaustion", which inexorably follows as long as the stressor is severe enough and applied for a sufficient length of time, since the "adaptation energy" or adaptability of a living being is always finite.

The resilience hypothesis, as defined by Flach (1990), is a theoretical construct that can be applied to the human response to stress and change:

> Psychobiological resilience is the efficient blending of psychological, biological, and environmental elements that permits human beings – and their families and, in fact, all humanly created organizations – to transit episodes of chaos necessarily associated with significant periods of stress and change successfully (p. 40)

According to Flach (1990), maintaining psychological homeostasis does not just involve restoring normal function; the new homeostatic condition after disruption should be different than before and represent a higher, more adaptable level of organization. In PTSD patients, the new homeostasis that has been created settled at a level of partial or nearly complete dysfunction. Obviously, these patients lack psychological resilience.

Boss (2006) defines resilience as "the ability to stretch (like elastic) or flex (like a suspension bridge) in response to the pressures and strains of everyday life" (p.48). When a crisis or a traumatic event occurs, resilience is defined as the ability to bounce back to a level of functioning equal to or greater than before the crisis. Resiliency also means thriving under adverse conditions and being able to maintain one's physical and emotional health and one's spirit for living life with joy. When a person is equipped with resiliency, he or she never really loses normal functioning but maintains a stable equilibrium no matter what. After a traumatic event, the majority of people are actually quite resilient, since only up to ca. 20 percent show symptoms of PTSD (Boss, 2006).

In sociology, the pioneer in resiliency (even though he did not use this term) was Aaron Antonovsky, who focused on a broader approach to viewing specific illnesses. He wanted to identify the psychological, social, and cultural resources that people use to resist illness. In his research, Antonovsky (1979, 1987) found that a sense of coherence was a major factor in determining how well patients would manage stress and stay healthy.

4.1.2 Hardiness

The concept of hardiness describes the cognitive beliefs and attitudes that enable an individual to deal with a difficult life crisis with adequate coping strategies so that he or she is resilient and transforms a potential disaster into an opportunity (Kobasa, 1979; Kobasa et al, 1982; Maddi, 2004).

According to existential psychologists, individuals continuously engage in a search for meaning in life that comprises a constant decision-making process. Regardless of the content, each decision involves choosing the unknown future or the familiar past. If people chose the future, this enables them to continue to elaborate life's meaning, but it invariably causes ontological anxiety. Thus, choosing the future requires courage. The construct of hardiness, which consists of the attitudes (1) commitment (versus alienation); (2) control (versus powerlessness); and (3) challenge (versus security), serves as an operationalization of existential courage (Maddi, 2004). In a longitudinal study by Maddi and Kobasa (1984) that assess the stress reactions of managers at Illinois Bell Telephone (IBT), the more the managers reported hardiness attitudes, the more they would respond to stressful circumstances with effective rather than ineffective coping, social interaction, and self-care behaviors, thereby increasing their likelihood of enhanced performance and health (Maddi & Kobasa, 1984).

A study by Kobasa (1979) with two groups of executives who had high degrees of stressful life events revealed that while one group suffered high stress without falling ill, the other group reported falling ill after their encounter with stressful life events. The proposition of the study was that individuals who experienced high degrees of stress without becoming sick had a personality structure that differentiated them from those who became sick. This personality disposition is best termed as hardiness. Hardy individuals are considered to possess three general characteristics: "(a) the belief that they can control or influence the events of their experience, (b) an ability to feel deeply involved in or committed to the activities of their lives, and (c) the anticipation of change as an exiting challenge to further development" (p. 3). A sense of control prompts actions aimed at transforming the events into something consistent with an ongoing life plan. When individuals feel they are in control, they are more likely to engage in active coping strategies. The relationship of committed people to themselves and to their environment involves being active and approaching the situation rather than resorting to passivity and avoidance. Moreover, committed people have a sense of purpose and find more meaning in the events. Challenge neutralizes the perceived severity of the events by making them appear stimulating rather than threatening. In terms of coping behaviors, challenge will lead to attempts to transform oneself and thereby grow rather than clinging to one's past existence. The results of Kobasa's (1979) study indicate that high stress/low illness executives show more hardiness as they have a stronger commitment to self, an attitude of vigorousness toward the environment, a sense of meaningfulness, and an internal locus of control compared to the high stress/high illness executives.

A few years later, Kobasa et al. (1982) tested the hypothesis that hardiness (commitment, control, and challenge) decreases the effects of stressful life events on physical health in a five-year prospective study. The hypothesis was supported, as the main effects on illness for both stressful life events and hardiness, as well as an interaction effect for these independent variables, were found. Therefore, hardiness has a health-preserving effect, especially when stressful life events increase.

This effect is not surprising given that, when personality dispositions such as hardiness operate at a psychological level, they have both cognitive appraisal as well as action aspects. At the appraisal level, they form the bases for experiencing stimuli in a certain way and ascribing a certain meaning to them. At the action level, the personality dispositions energize a particular set of activities that are experienced as appropriate. These activities are often referred to as *coping* (see Coping Strategies, 4.2). In this way, personality dispositions can have a buffering effect on stressful events, especially those that have the cognitive appraisal effect of rendering the events less meaningless, overwhelming, and undesirable, and the action effect of initiating coping that is oriented toward improving the situation (Kobasa et al., 1982).

In a study by King, King, Fairbank, Keane, and Adams (1998), structural equation modeling procedures were used to examine the relationships between resilience-recovery factors (hardiness, post-war social support, and additional negative life events after the war) and Vietnam War veterans' PTSD symptoms. The results indicate strong mediation effects of these factors. War zone stressors and stressful life events in the post-war period both decrease the level of hardiness. Hardiness in turn leads to an increase of structural and functional social support and to a decrease of PTSD symptoms. Additional stressful life events also decrease functional social support, as they deplete social resources. Nevertheless, more hardy individuals are better able to mobilize social resources than less hardy individuals in times of stress.

4.1.3 Sense of Coherence

Antonovsky's (1987) main focus of attention is the intriguing phenomenon that most organisms emerge from encounters with stressors quite unscathed. His focus is on the deviant case, for example, "Who are the type As who do not get coronary heart disease? Who are the smokers who do not get lung cancer?" (p. 11). Antonovsky's model of health and disease is a salutogenetic model. Therefore, instead of viewing a person as either sick or healthy, he or she can be placed somewhere along a continuum between health and ease/dis-ease. In this way, the focus is on "coping rather than stressors, salutary factors rather than risk factors, survivors rather than the defeated, the invulnerable rather than the defeated" (Antonovsky, 1992, p.34). However, the salutogenetic orientation is not merely the opposite of the pathogenic orientation, but is radically different. Underneath the model lies the fundamental assumption of heterostasis as opposed to homeostasis. In this way, the prototypical characteristic of the living organism is one of heterostasis, disorder, and pressure toward increasing complexity. The conventional pathogenic assumption is that an organism strives to maintain homeostasis, which is regulated once it is disrupted. Along these lines, hypotheses are formulated focusing on specific diseases and risk factors. The stressor is the main focus. In contrast, the salutogenetic orientation focuses on factors promoting movement toward the healthy end of the continuum. These often differ from reversed risk factors. Consequently, the main focus is on coping and determining what factors not only act as buffers, but also contribute directly to health. Stressors are viewed not necessarily as pathological but, quite possibly as salutary (Antonovsky, 1987). Just being alive is potentially pathogenic, and in the long run, everyone is a terminal case. However, the stressor does not determine the outcome (except perhaps for the most extreme circumstances such as a concentration camp where survival is rare and arbitrary). As Almedom (2005) states: "Positive and negative aftermaths, 'recovery' and 'chronic trauma', respectively, are essentially two sides of the same coin. What is it in individuals and/or their social environment that determines which side of the coin is facing up when adversity strikes" (Almedom, 2005, p. 253)?

As an answer to the salutogenetic question, Antonovsky came up with the construct of "sense of coherence" (SOC), which is an important determinant of maintaining one's position on the health ease/dis-ease continuum and of movement towards the healthy end. The SOC acts as a buffer to mitigate the effects of potentially stressful events or directly promotes health. According to Antonovsky, generalized resistance resources such as high social class, self-esteem, social support, and cultural stability, create life experiences characterized by consistency, participation in shaping the outcome, and an underload-overload balance, which reinforce a strong SOC. Thus, a crystallized, integrated view of the world is generated. In the same way, properties such as low self-esteem, isolation, low social class, or cultural instability provide contrary experiences and thus function as generalized resistance deficits (Antonovsky, 1987,

1992), and lead to a weak SOC. The balance of the experiences provided by the generalized resistance resources and deficits in one's life leads to one's location on the SOC continuum.

The SOC is a characteristic of a person. Nevertheless, Antonovsky refrains from calling it a personality type, preferring the term "dispositional orientation" (Antonovsky, 1992, p.37). The author holds the view that during young adulthood, at about 30 years of age, the SOC becomes stable unless patterns of experiences in life change drastically.

The SOC is characterized by three components:

(1) *Comprehensibility*: This refers to the extent to which one perceives the environment "as making cognitive sense, as information that is ordered, consistent, structured, and clear, rather than as noise – chaotic, disordered, random, accidental, inexplicable" (pp.16/17). Future events are predictable for someone with a strong SOC, and when the events happen unexpectedly, they are at least orderable and explicable.

(2) *Manageability*: This is defined as the "extent to which one perceives that resources are at one's disposal which are adequate to meet the demands posed by the stimuli that bombard one" (p. 17). These resources might be under one's one control or under the control of legitimate others (e.g. one's spouse, friends, colleagues, God, history, party leader, a trusted physician). This enables individuals to cope with unfortunate events that happen.

(3) *Meaningfulness*: This component represents a motivational element. People with a strong SOC speak of areas of life that are important to them and that "make sense" not only in the cognitive, but also in the emotional sense. The meaningfulness component refers to the "extent to which one feels that life makes sense emotionally, that at least some of the problems and demands posed by living are worth investing energy in, are worthy of commitment and engagement, are challenges that are 'welcome' rather than burdens that one would rather do without" (Antonovsky, 1987, p. 18).

Antonovsky (1987) finally defines SOC as follows:

> The sense of coherence is a global orientation that expresses the extent to which one has a pervasive, enduring though dynamic feeling of confidence that (1) the stimuli deriving from one's internal and external environments in the course of living are structured, predictable, and explicable; (2) the resources are available to one to meet the demands posed by these stimuli; and (3) these demands are challenges, worthy of investment and engagement (p. 19).

Antonovsky suggests that although all three components are necessary, not all are of equal centrality. The most crucial one seems to be the motivational component of meaningfulness. If one does not have meaningfulness, comprehensibility and manageability will only be temporary. Next in importance is comprehensibility, as one needs understanding in order to cope. Nevertheless, manageability is also important. If a person does not believe that resources are at his or her disposal, meaningfulness will become less and coping efforts are weakened. Successful coping therefore depends on SOC as a whole.

Antonovsky has attempted to answer the question: "How does SOC promote health?" First of all, according to Lazarus and Folkman (1984); "Psychological stress is a particular relationship between the person and the environment that is appraised by the person as taxing or exceeding his or her resources and endangering his or her well-being" (p.19). In Antonovsky's view, a person with a strong SOC is more likely to appraise a stimulus as a non-stressor in the first stage of appraisal. If, however, a stimulus has been defined as a stressor, persons with a strong SOC are more likely to define the stressor as benign or welcomed in the second stage of appraisal. These persons are confident that they will manage well and the emotions aroused are salutary. In the third stage of appraisal in the coping process, the problem is defined. When a person has the generalized expectation that stressors are meaningful and comprehensible, this forms the motivational and cognitive basis for coping and preventing the transformation of tension into stress. The person with a weak SOC, however, is confused and devoid of the desire to cope and tends to give up at the outset. A person with a strong SOC chooses the particular coping strategy from among the variety of potential resources that seems most appropriate to deal with the given stressor. In this way, he or she is flexible in employing different strategies and uses feedback to correct the coping strategy, if necessary. A person with a weak SOC might focus on the emotional parameters, on handling the unpleasant emotions caused by the stressor, whereas a person with a strong SOC will tend to focus on the instrumental parameters of the problem and mobilize resources to deal with it (Antonovsky, 1987). When coping is successful and the tension is resolved, the very experience of successful coping is emotionally gratifying and will have salutary physiological consequences. Moreover, when one considers psychoneuroimmunology, a very direct relationship may exist between SOC and health: the perception of the world of stimuli as comprehensible, manageable, and meaningful might activate the brain to send messages to other bodily systems that maintain homeostasis (Antonovsky, 1992).

When comparing the SOC with other salutogenetic-oriented models such as the concept of hardiness (Kobasa, 1979), Antonovsky's view is that "Whatever words are used, it seems to me that we are all talking about exactly the same thing" (p.49). Meaningfulness implies that some events that happen are seen as challenges worthy of commitment. Kobasa writes of commitment to self and vigorous involvement in one's life. Comprehensibility can be compared to Kobasa's challenge, as change is normative and the change as well as the unexpected and unpredictable are ideally viewed as an opportunity to turn them into something coherent. Manageability is congruent with control, the only difference being that Kobasa mentions internal control, whereas Antonovsky broadens the spectrum of control to include control exercised by legitimate others. Moreover, persons who hold a strong coping capability may be more resilient in the face of challenging and stressful events because they are more likely to possess the skills and abilities required to resolve difficult circumstances (Antonovsky, 1987).

SOC and Psychophysical Health

In a study with Israeli retirees, Antonovsky (1992) hypothesized that the greater the stressor load, the more important the role of salutary factors. This longitudinal study of the developmental transition of retirement from work was the first empirical study showing the contribution of SOC to health. The study aimed to clarify the characteristics of those people who coped successfully with the stressor, as well as the conditions that facilitated such coping. The results indicate that SOC has both direct and indirect consequences for health status, as generalized resistance resources such as attitudes toward retirement and activity level constitute a channel through which SOC affects health status.

Several studies have demonstrated a positive correlation between the existence of a sense of coherence and psychophysical health and adequate coping (Höge, 2005; Groß & Pöhlmann, 2004; Schüssler, 2003; Sek & Tasikowski, 2002).

In a path analysis model, Höge (2005) demonstrated that relationships exist between organizational resources of mobile nursing services, the perceived fairness of the interaction between the organization and the nurses, the sense of coherence of the nurses, and their psychological health. The results indicate that perceived fairness increases both an individual's sense of coherence and psychological health.

Furthermore, Groß and Pöhlmann (2004) investigated PTSD and burnout symptoms in paramedics. They found that emotion-focused and avoidant coping strategies increased symptom levels, although people with a high degree of sense of coherence did not engage in these inefficient coping strategies but used more efficient ones.

A relationship between a strong sense of coherence and good health status was also found in a study with Swiss mountain guides (Schüssler, 2003). Another study aimed at measuring coping with stress along the salutogenesis model revealed a correlation between a strong sense of coherence and a tendency to perceive stressful events as a challenge rather than a threat or loss. A strong sense of coherence also instigated the choice of active rather than avoidant coping strategies (Sek & Tasikowski, 2002).

In a study with Israeli undergraduates, Cohen, Ben-Zur, and Rosenfeld (2008) tested SOC, coping strategies, and test anxiety as predictors of test performance. The short 13-item SOC and a 30-item Hebrew version of the COPE Scale, based on the scale developed by Carver, Scheier, and Weintraub (1989), were used. SOC was negatively related to emotion-focused coping, avoidance, and test anxiety and positively correlated with problem-focused coping. Emotion-focused coping and avoidance were positively related to test anxiety. Higher test anxiety led to a lower test performance. However, there was no direct effect of SOC coping on test performance.

The SOC model appears to be valid across many cultures. A recent follow-up study with Japanese university students revealed that SOC had an effect on well-being after two years (Togari, Yamazaki, Takayama, Yamaki & Nakayama, 2008). In a multiple-regression analysis, high baseline-SOC and later SOC change were found to be predictors of high physical and mental well-being. The results differed between genders, with SOC influencing successful coping with developmental tasks more in males. It is speculated that the issue of choosing an appropriate career is of more relevance to young males in Japan and can therefore be an extremely big stressor for them.

Similar results were obtained for a Finnish sample. In a population-based study by Konttinen, Haukkala, and Uutela (2008), SOC, depressive symptoms, and anxiety, as well as their relationships with health, were compared. Strong negative correlations between SOC and measures of depression and anxiety were observed. However, the question arises whether the SOC scale inversely measures the other two constructs. To answer this question, the discriminant validity of the three measures was examined by comparing their associations with health indicators and behaviors. Higher SOC scores and lower depression and anxiety scores were related to better self-rated health, fewer visits to a doctor, and less somatic and psychological symptoms in both genders. The intercorrelations among the SOC and health variables were stronger among the respondents in the lowest than highest SOC tertile. Apparently, only a low and not a high SOC predicted health. Confirmatory factor analysis (with maximum likelihood estimation) was also used to compare the applicability of a one-factor and correlated three-factor model of the SOC items to the present data. A comparison of the SOC and the BDI items revealed some content overlap. Four meaningfulness items and one manageability item (1, 4, 7, 10, and 12) (see Appendix) could be interpreted as reflecting symptoms measured with the BDI, and two comprehensibility items (8 and 9) seemed to assess negative emotional experience in general. In addition, two manageability items and one comprehensibility item (2, 3, and 5) were more likely to measure interpersonal trust/mistrust than the manageability or comprehensibility dimensions of SOC. It is therefore not surprising that factor analyses conducted on their data are rather ambiguous. The authors decided to use the one-factor model, as the factors were highly correlated in the three-factor solution. The authors conclude that even though an overlap exists between the SOC scale and measures of depression and anxiety, the current psychometric methods appear able to separate highly correlated measures, and these small differences could have value in some research settings.

Another study with an Asian sample investigated the role of SOC in relation to depressive symptoms for Taiwanese family caregivers of cancer patients (Tang & Li, 2008). It was confirmed that among the independent variables investigated in this study, variations in SOC were important in explaining the variation of depressive symptoms. The strength of SOC in family caregivers was moderately strong [mean (S.D.) = 59.81 (15.88)]. Potential correlates of depressive symptoms were categorized as: (1) characteristics of the family caregivers and cancer patients, (2) care giving demands: objective care giving loads, (3) appraisal of the care giving situation, including confidence in end-of-life care giving, and subjective care-giving burden, and (4) available resources: personal coping capability (the SOC). In the final step of the multiple regression analysis, the SOC of caregivers increased the explained variation in depressive symptoms by 21.5 percent. The results indicate that the extent of the experienced stressful event, in this case providing care for a terminally ill relative, does not only depend on the demands of the situation, but also on the relationship between the demands and the available social and psychological resources, as perceived by the individual.

One might argue that there is a reciprocal relationship between SOC and health factors. Poor health might be another stressor with which one finds difficulty in coping, and this experience reinforces the belief that life is incomprehensible, unmanageable, and meaningless. However, Antonovsky's (1992) retirement study showed that SOC better predicts health status a year later than health status predicts SOC.

SOC and Trauma

Which direction along the continuum between healthy and sick a person moves toward after an encounter with a trauma depends on the balance between the severity of the stressor and their SOC (von der Heyden, 2006). However, a study with traffic accident victims by Frommberger, Stieglitz, Straub, Nyberg,

Schlickwei, Kuner, and Berger (1999) showed that cognitive and emotional reactions to a motor vehicle accident strongly depended on the people's sense of coherence, independent of the severity of the accident. Accident victims who fulfilled the criteria for PTSD or subsyndromal PTSD after the accident revealed a significant lower SOC score than victims without any psychiatric diagnosis. Altogether, the SOC total score correlated negatively with (a) post-traumatic pathology, (b) psychological disorders, and (c) anxious cognitions. The personality trait of neuroticism correlated negatively and extraversion and frustration tolerance positively with SOC.

4.2 Coping Strategies

Coping is usually defined as behavior targeted at protecting people from being psychologically harmed from stressful events. It plays a mediating role between stress and psychological, physical, and social well-being. In an early study on the structure of coping, Pearlin and Schooler (1978) suggested that this protective function can operate in three ways: (a) by modifying the outer circumstances that cause problems, (b) by modifying the meaning of experiences from problematic to neutral, and (c) by regulating one's emotional reactions to the problems. Moreover, these authors distinguish between social resources, psychological resources, and specific coping responses. In this context, resources refer to what is available to people to develop their coping repertoires, and coping responses are what people actually do.

Rather than conceptualizing coping in terms of defensive ego processes (Haan, 1977; Vaillant, 1977) or as personality traits (Lazarus, Averill, & Opton Jr., 1974) or in terms of the demands of specific situations (Moos, 1977), in 1980, Folkman and Lazarus conceptualized coping as an interaction between person and environment. Their definition of coping is process-oriented rather than trait-oriented. The underlying theoretical framework is transactional in that the person and the environment are in an ongoing reciprocal relationship mediated by two processes: appraisal and coping. Coping efforts are made in response to stress appraisals. Appraisal is the cognitive process through which an event is evaluated in terms of harm/loss, threat, or challenge (primary appraisal) and which coping resources are available (secondary appraisal). The changed person-environment relationship leads to new appraisals (or reappraisals), which in turn elicit further coping efforts. In 1984, Lazarus and Folkman gave the often-cited definition of coping as "constantly changing cognitive and behavioral efforts to manage specific external and/or internal demands that are appraised as taxing or exceeding the resources of the person" (p.141).

In a study exploring the ways in which 100 community residents cope with the stressful events of daily life, Folkman and Lazarus (1980) analyzed two coping functions: problem-focused and emotion-focused coping. The former is exercised to manage or alter the person-environment relationship, that is, the source of stress, and emotion-focused coping is used to regulate stressful emotions. The influences of factors such as the context of an event, who is involved, how it is appraised, age, and gender on coping were examined. The results indicate that the context and appraisal of the event were the most influential factors. Work contexts favored problem-focused coping, whereas health contexts favored emotion-focused coping. Situations in which people thought something constructive could be done or that were appraised as requiring more information favored problem-focused coping, whereas situations in which nothing could be done and had to be accepted favored emotion-focused coping. There were no age or gender effects in regard to problem-focused coping. Men employ more problem-focused coping in work situations and situations that need to be accepted and require more information. No gender differences in emotion-focused coping were found.

4.2.1 Coping Efficacy

Pearlin and Schooler's (1978) criterion for coping efficacy is how well a coping response prevents the life-strains people experience, which result in emotional stress. The authors asked the research question whether some coping responses were more effective than others or whether the sheer scope and variety of coping responses would have a positive effect. In a study of the general population of an urbanized area of Chicago, they found that the effect of any single coping mechanism was rather modest. In occupational and economic areas, the most effective types of coping involve the manipulation of goals and values. However, in marriage and parenthood roles, commitment and involvement were most effective and avoidance and withdrawal were least effective. A wide range of coping responses is definitely more effective than a single coping response, regardless of its efficacy.

Folkman and Lazarus (1980) discovered that both problem-focused and emotion-focused coping were used in 98 percent of the stressful episodes mentioned by the participants. According to the authors, it does not make sense to speak about the efficacy of either problem-focused or defensive mechanisms, as researchers had done previously, given that both functions are usually involved. Moreover, intra-individual analyses showed that people are more variable than consistent in their coping patterns. According to Lazarus and Folkman (1984), effective coping is characterized by problem- and emotion-focused coping working in a complementary fashion without impeding each other. Coping strategies are not inherently good or bad. A strategy that is effective in one situation can be ineffective in another, and vice versa. Three basic kinds of adaptational outcomes are affected by coping processes: functioning in work and social living, morale or life satisfaction, and somatic health (Lazarus & Folkman, 1984). Appraisals need to match the flow of events. A high level of perceived threat can lead to more emotion-focused coping. If the situation is uncontrollable, emotion-focused strategies such as rearranging priorities or avoidant behavior might be highly functional. In another more recent study, it was shown that for a sample of Palestinian political ex-prisoners and a non-prisoner control group, none of the dispositional and situational coping responses were effective in protecting against the negative impact of trauma. However, the main effect results revealed that a high level of active and a low level of emotion-focused dispositional coping was associated with lower levels of psychological distress. In contrast, situational responses of avoidance and denial were associated with lower levels of PTSD, depressive, and somatoforms symptoms (Punamäki, Salo, Komproe, Qouta, El-Masri, & De Jong, 2008). Similarly, a prospective study with undergraduates who were indirectly exposed to the 9/11 World Trade Center attacks showed that at both one and three months after the attack, higher levels of pre-9/11 emotion-focused coping predicted higher levels of PTSD symptomology (Baschnagel, Gudmundsdottir, Hawk, Gayle, & Beck, 2009).

Besides the focus approach (problem-focused versus emotion-focused), researchers have used other conceptual approaches to classify coping responses, for example, according to the method (cognitive versus behavioral), or to the orientation (approach versus avoidance). Several studies in Western cultures have shown that people who tend to engage more in approach coping and less in avoidance coping show lower levels of PTSD symptomology. Avoiding feelings, thoughts and actions associated with the traumatic situation does not lead to successful coping with the trauma (Solomon et al., 1988; Dunmore, Clark & Ehlers, 2001) An avoidant coping style is generally considered as predicting a higher level of PTSD, for example, in victims of domestic violence (Krause, Kaltman, Goodman, & Dutton, 2008) or in victims of Hurricane Katrina (Glass, Flory, Hankin, Kloos, & Turecki, 2009). In another recent study with Hurricane Katrina evacuees, Sprang and LaJoic (2009) found avoidance coping to be a partial mediator between the degree of exposure to the hurricane and the onset of PTSD symptoms. The more avoidant coping behavior the respondents showed, the higher the likelihood of experiencing threshold levels of PTSD one year after the disaster.

A longitudinal study with American Vietnam War veterans diagnosed with PTSD by Tiet, Rosen, Cavela, Moos, and Finney (2006) investigated the association between approach coping and better functioning outcomes and the reciprocal relationships between coping and posttraumatic stress disorder (PTSD) symptoms in patients diagnosed with PTSD were investigated. Analyses using structural equation modeling techniques showed that more approach coping predicted better family and social functioning. Cognitive avoidance coping, on the other hand, predicted more PTSD symptoms. A high level of PTSD symptoms predicted more approach coping and more behavioral avoidance coping, which may be due to the fact that patients with more PTSD symptoms are expected to experience more stressors, which elicit more coping responses. The authors' conclusion is that approach coping may enable patients with chronic PTSD to establish and maintain better relationships with family and friends, despite continuing PTSD.

4.2.2 PTSD and Coping after Natural Disasters

Gerrity and Steinglass' (1994) interviews with families who lost their homes due to a flood revealed three positive coping strategies. These were (1) the reordering of priorities, especially the redefining of material possessions as having less meaning than in the past; (2) personal and fairly constant immersion in recovery activities, directed toward safeguarding what could be saved, and letting go of what could not; and (3) the development of a new understanding of the meaning or purpose of life; for many, a new relationship with God or a realignment with one's family or social world. Dysfunctional coping strategies were found to be (1) displacement of emotion, as in diverting all grief and painful emotions onto those they deemed responsible for the disaster; (2) avoidance of social contact; and (3) family conflict.

Ten different coping strategies used by fourth-grade children affected by Hurricane Floyd were assessed by Russoniello et al. (2002). The coping strategies significantly related to PTSD symptoms were social withdrawal, self-criticism, blaming others, problem solving and emotional regulation. Wishful thinking (hope) was the most frequently used coping strategy and the strategy least associated with PTSD symptoms. Other strategies not significantly associated were distraction, social support, cognitive restructuring, and resignation (accepting the reality of the event). The girls were significantly more likely to use social support than the boys. In addition, those children whose homes were flooded were significantly more likely to use emotional regulation and social support than those who were not flooded.

Three months after the devastating Hurricane Andrew in Florida, children reported using wishful thinking coping most frequently, followed by positive coping, social withdrawal, and blame-anger, respectively (Vernberg, La Greca, Silverman, & Prinstein, 1996). All four coping variables were positively correlated with the overall PTSD symptoms. The reason for this finding may be that high levels of distress following a novel, intense, ongoing stressor such as a natural disaster may initially elicit a variety of coping strategies, both positive and negative.

In a study with a community sample of self-defined trauma survivors, Goldenberg and Matheson (2005) investigated the mediating role of coping between inner representations and PTSD symptomology. The authors combined problem-focused coping and seeking social support to "active coping" and self-blame, avoidance, and wishful thinking to "passive coping". More negative inner representations (or appraisals) were related to a decreased tendency to use active coping and an increase of the use of passive coping strategies. The use of active coping strategies was not related to the level of PTSD symptoms, although the endorsement of passive coping was strongly positively related to the level of PTSD symptoms.

It is very likely that the relationships between coping and traumatic stress symptoms are reciprocal and dynamic. Spurrell and McFarlane (1993) investigated the role of coping in the onset of PTSD symptoms after a bushfire disaster in Australia. In contrast to other studies, the use of all coping strategies was found to be associated with the presence of PTSD rather than the absence of symptoms. The DSM-III disordered participants used all the coping behaviors to a greater degree, that means problem-focused as well as emotion-focused coping. These findings suggest that coping in this sense represents a psychological process used to limit the distress caused by symptoms as well as to overcome adverse situations. Obviously, the relationship between events, coping, and disorder is quite complex. This study only suggests an association between coping behaviors and psychiatric disorders and the direction of causality could not be specified. Moreover, the authors argue that some of the items of the Ways of Coping Questionnaire (Folkman & Lazarus, 1985) belong more to a measure of appraisal than coping effort. In fact, Horowitz et al. (1979) attempted to separate these processes by studying participants' cognitive and affective appraisal of life events to develop the Impact of Event Scale with the two dimensions of cognitive processes intrusion and avoidance. Another explanation of the authors' contradictory findings is that participants reported the efforts to cope with the ongoing reactivation of the trauma. As the event is experienced afresh each time, problem-focused and emotion-focuses strategies are agein employed to deal with the emotional reaction. The authors' findings might also be determined by the extreme nature of the disaster, which presents an unsolvable situation (Spurrell & McFarlane, 1993).

4.2.3 Cultural Aspects of Coping

One of the most important neglected contexts of coping is culture. As coping depends on the dynamic interaction between person and environment (Lazarus & Folkman, 1984), Chun et al. (2006) believe that culture is a fundamental context that shapes both the individual and the environment. Culture is defined as a highly complex, continually changing system of meaning that is learned, shared, transmitted, and altered from one generation to the next (Triandis, 1995). This system of meaning includes the norms, beliefs, and values that provide directions for behavior. Among the various cultural values that have been identified and studied, individualism and collectivism are the most widely studied, as they are able to meaningfully differentiate cultures. The constructs of individualism and collectivism are comprised of a set of values, attitudes, and behaviors that vary in the priority placed on the self versus the ingroup. In more individualistically-oriented cultures, the self is the central unit of society. Accordingly, individual rights, a concern for oneself and immediate family, and personal autonomy and self-fulfillment are emphasized. On the contrary, in more collectivistically-oriented cultures, the in-group forms the central unit of society and binds individuals to its needs, goals, and fate. Consequently, duty and obligations to the in-group, interdependence on other individuals within the group, and the fulfillment of social roles are emphasized (Hofstede, 1980). Culture also influences the ways in which individuals define themselves in relation to their social context. Individualistic cultures tend to cultivate an independent self-construal that defines self as a separate and independent entity with well-defined boundaries and in terms of abstract and general traits. Collectivistic cultures, however, tend to cultivate an interdependent self-construal that defines self in relation to other people, with overlapping interpersonal boundaries and in terms of social roles and situations (Markus & Kitayama, 1991).

In a cross-national study by Hofstede (1980) with people who worked for a multinational corporation in 39 countries, respondents in the United States were shown to be most individualistic and respondents in Japan and East Asian nations were the least individualistic. Moreover, Oyserman, Coon, and Kemmelmeier's meta-analysis (2002) confirmed these findings in that people in the United States, especially European Americans, tend to be more individualistic and less collectivistic than people in non-Western and/or developing nations.

Culture influences the environmental system to a great extent and thus the dominant aspects of the social climate. For example, in individualistic societies such as the United States, the social climate tends to value and protect individuals' rights to autonomy and independence. The social norm is to individuate from family at the end of adolescence; thus cohesion with family and group may be played down. In contrast, the social climate of societies that are more collectivistic tends to promote social conformity and interdependence. Pursuit of individual autonomy is often viewed as selfish and a betrayal of the in-group. Instead, personal sacrifice for the larger community is regarded as an indication of maturity and strong character. The environmental system also determines the pressures and demands on the individual. In individualistic societies, many of the ongoing stressors may stem from the pressure to be autonomous and independent. These pressures occur in early adulthood, when a person may not be ready to individuate or handle all responsibilities alone. The emphasis on independence and autonomy may also be associated with fewer available resources. In contrast, in collectivistic societies, there is greater pressure to remain interdependent on the in-groups and to meet their demands, sometimes at the expense of the individual's own welfare (Chun et al., 2006).

Moreover, individuals from collectivistic cultures tend to make more external attributions. Kawanishi (1995) found that Japanese college students were more likely than Euro-American students to make attributions to external chance factors. She suggested that this attribution pattern was observed because of the Japanese common belief in luck, a type of external locus of control, which has been associated with collectivism. In addition to having a more external locus of control, people from collectivistic cultures tend to be more field-dependent and more affected by environmental demands (Markus and Kitayama, 1991). For individuals with a more interdependent self-construal, the target of control is likely to be the self because the individual needs to fit into the group and protect it. Therefore, these individuals are bound to exercise control by trying to change their mind, emotions, thoughts, and behaviors as opposed to trying to change external factors (Chun et al., 2006).

From this perspective, it is not surprising that culture influences coping goals and selection of coping strategies. In collectivistic cultures, the immediate goal of the individual is not necessarily the reduction of his or her own distress, but the well-being of the in-group. Past research has held the individualistic view that individuals set coping goals that primarily address their own needs, as in the immediate reduction of their own psychological distress (self-focused coping goals), rather than goals focused on meeting the needs of other people (other-focused coping goals). Individuals from collectivistic cultures whose interdependent self-construal embraces other members of the in-group as part of the self make the welfare of the in-group an integral determinant of the welfare of the individual. When other-focused coping goals require some amount of self-sacrifice, this may result in an immediate increase in distress, rather than the decrease of distress that researchers often use as evidence of effective coping. This may explain the frequent findings that Asians and Asian Americans tend to use seemingly ineffective coping strategies that do not reduce their psychological distress (Bjorck, Cuthbertson, Thurman, & Lee, 2001; Essau & Trommsdorff, 1996; Lee & Lui, 2001). The coping strategies known to be effective based on research that use symptom relief as a measure of successful outcome may not necessarily be those that help other types of desired coping goals. For example, the goal of conflict resolution would not necessarily be to remove the conflict or barrier to the desired outcome, but rather to manage the conflict. In a collectivistic cultural context, compromise and accommodation are regarded as an indication of maturity and tactfulness (Chun et al., 2006).

Coping strategies also vary according to the culture. Coping strategies that confront and modify external stressors (e.g. behavioral or approach-focused coping strategies) are expected to be more common in individualistic cultures, whereas coping strategies that avoid external stressors and modify internal psychological states (e.g. cognitive or avoidance-focused coping strategies) are expected to be more common in collectivistic cultures. The reason is that persons from individualistic cultures have a greater sense of internal locus of control and cognitive styles oriented toward dispositional causal attribution, especially for stressors. Consequently, these individuals are more prone to attempt to influence the external environment in order to achieve their desired coping goals. In contrast, individuals from collectivistic cultures, who have more external locus of control, greater secondary control, and a greater tendency to attribute stressors to bad luck, are expected to have a greater desire to control their internal states in order to achieve their coping goals. To the extent that these individuals believe that they do not have control over the outcome and are at the mercy of powerful forces, luck or fate, they will most likely feel helpless and thus reliant on passive or avoidant coping strategies (Chun et al., 2006). Cultural beliefs in fate (also known as *karma* in Asian cultures), may also aid in absolving an individual of an undue sense of failure or incompetence and in warding off depression (Aldwin, 2007). Several studies confirmed that adults and children of collectivistic cultures, such as Korean Americans, Malays, and Ghanians, were more likely to use passive coping; whereas adults and children of individualistic cultures, such as Euro-Americans and Germans, were more likely to use action-oriented and problem-focused coping. However, people rely on multiple coping strategies and individual and situational factors can sometimes override cultural influences (Bjorck et al., 2001; Essau & Trommsdorff, 1996). Finally, for stressed collectivistic individuals, choosing appropriate coping strategies appears to involve finding a balance between taking care of the needs of the individual, maintaining the well-being of the in-group, and protecting the relationship between the individual and the in-group (Chun et al., 2006).

In studies conducted in Western industrialized countries, empirical findings suggest that approach coping is constructive and adaptive, whereas avoidance coping is passive and maladaptive. However, when a stressor is uncontrollable or unavoidable, approach coping strategies may be inefficient. In addition, palliative or avoidance strategies may be more effective in some cultural contexts where the norm is to "fit in" with the social and physical environment. In Eastern collectivistic cultures, the worldview is more holistic and the separation between person and environment is considered artificial and meaningless. When problems arise between the self and the environment, the cause is perceived to be neither within the person nor in the external world, and a "mature" person would take actions to control the self to make it fit better with the environment. Thus, coping strategies focusing on exercising secondary control, i.e. controlling one's own internal states and behaviors, are deemed desirable and may be more effective in achieving coping goals. Such coping efforts are neither passive nor avoidant: in reality, they require intense effort and concentration on the target of control, the mind and behaviors of the self (Chun et al., 2006).

The assumption that active or approach-focused coping strategies are more adaptive than avoidant coping strategies has been challenged in several studies. For example, Chang (2001) investigated the normative coping strategies of Euro-American and Asian American college students and their relationships to positive and negative psychological outcomes. Compared to Euro-American students, Asian American students reported a greater use of problem avoidance and social withdrawal. The use of these avoidant strategies was associated with less life satisfaction and more depression in Euro-Americans, but it was not associated with either of these outcomes in Asian Americans. Likewise, Yoshihama (2002) found that the perceived effectiveness of passive coping strategies was associated with lower psychological distress in the more collectivistic Japanese-American women born in Japan, whereas the perceived effectiveness of active coping strategies was associated with lower distress in their more individualistic counterparts born in the United States. Furthermore, active coping appeared to have deleterious effects on the psychological well-being of the Japan-born women. In a sample of Ethiopian internally displaced persons, avoidance-oriented coping was significantly negatively correlated with mental distress and significantly positively correlated with average quality of life. Emotion-oriented coping, on the other hand, was significantly positively correlated with mental distress and significantly negatively correlated with average quality of life. Task-oriented coping was also beneficial, as it decreased mental distress, but only if the trauma was severe. Otherwise, there was no significant difference in the corresponding mental distress values (Araya, Chotai, Komproe, and de Jong, 2007).

In a study with Jewish-Israelis and Arab-Israelis examining different ways of coping with terrorist attacks, it was found that adaptation (using a variety of problem-solving strategies) and collaborative coping (employing social action) were predictors of PTSD, while the more emotion-focused ways of coping (acceptance, avoidance, and personal care) were not. Acceptance coping was best predicted by Arab-Israeli ethnicity, female gender, greater religiosity, and lower education (Somer, Maguen, Or-Chen, & Litz, 2007).

In a study with tsunami survivors in Azheekkal, Kerala, by Sunidharan (2005), certain demographic factors and coping strategies were investigated one to three months after the tsunami to determine the risk factors for developing anxiety and depression. Participants reported using distraction coping strategies most frequently, followed by problem solving, faith, acceptance, and denial. Social support seeking strategy was least reported. The tsunami victims who had lost at least one family member were using comparatively less coping strategies, and significant differences were observed in their coping strategies of problem solving, distraction, denial, and social support. In addition, participants suffering from chronic illness used the coping strategy of acceptance more than their counterparts. Coping with substance use led to a temporary decrease in anxiety and depression scores. To overcome anxiety, the coping strategy of distraction was the most successful, as there was a significantly negative correlation between distraction and anxiety score. Emotion-focused strategies were more frequently used (denial, distraction, faith, acceptance) than problem-focused (problem solving) ones. That means the survivors preferred activities that distracted them from thinking about the traumatic experience. Men scored higher on most coping strategies than women. Young adults also reported using more problem-focused coping than other age groups.

4.3 Social Support

In the scientific literature, the concept of social support is very diverse and the operationalization also varies considerably. The operationalization of social support can be objective (e.g. the size of the social network, marital status, number and availability of social resources) or subjective (perceived social support, the sense of being accepted). Social support is the result of cognitive-emotional processing of current and past interactions, through which the individual is receiving or has received support in achieving their personal goals or coping with challenges (Fydrich and Sommer, 2003). Perceived social support, which refers to the perception that help will be available when needed, is an empirically and theoretically distinct construct from received support, which is defined by help that has already occurred during times of need. For Fydrich and Sommer (2003), the subjective rather than objective component is crucial in determining social support. For example, even if someone is married, he or she might not feel supported by his or her spouse. Support given with the best intentions may also not be appreciated by the receiver.

According to Fydrich and Sommer (2003), the contents of social support can be divided into (a) emotional support (e.g. experiencing affection, recognition, closeness, trust, acceptance), (b) tangible support (material, concrete support), and (c) social integration (belonging to a social group, joint activities with others, joint value system, stability of relationships). Their own assessment method measures subjectively perceived social support as well as the quality of the social network using the Social Support Questionnaire (Fragebogen zur Sozialen Unterstützung, F-SozU-S54 or F-SozU-K22) (Sommer & Fydrich, 1989; Fydrich, Sommer, Menzel, & Höll, 1987).

4.3.1 Empirical Evidence

Social support can influence mental and physical health in two ways: firstly, social support can directly lead to greater well-being (main effect); and secondly, it can modify the effect of stressors on well-being (buffer effect). The main effect can be explained as follows: social support directly fulfills basic social needs such as belonging, human contact, socializing, and sense of belonging, and thus it creates a sense of well-being (Sommer & Fydrich, 1989). According to Lazarus and Folkman's (1984) transactional stress model, the buffer effect can be explained as follows: perceived social support has a positive effect on the primary and secondary appraisal of a stressor, since people who have adequate social support believe that they have access to a larger number of coping strategies.

However, the effects remain unpredictable, being shaped by personal, social, and contextual factors in ways that are still unclear. Review papers summarizing the buffering and/or direct effect have reported these difficulties in demonstrating consistent effects. Some effects are only observed under certain conditions. This may be due to the ambiguity of the construct, the low quality of measures, lack of sufficient sample size, and the diversity of stressor and distress variables (Vaux, 1988). Sommer and Fydrich (1989) also remind people that they need to apply caution when comparing the results of studies investigating social support, as many studies face methodological and conceptual difficulties. For example, most studies are retrospective self-report studies, and memories can be biased. Moreover, the conceptualization of social support varies considerably from study to study. Sometimes, social network size or the frequency of social contacts is measured, whereas other times the focus is the subjectively perceived quality of social support. Therefore, it is difficult to compare studies. Even if the correlations between social support and psychological or physical symptoms are significant, they are often not very high ($r < 0.30$) (Sommer & Fydrich, 1989).

Social support interactions are more complex than assumed and are not always linear and beneficial, according to Cruza-Guet, Spokane, Brown, and Szapocznik (2008). Moreover, Joseph, Williams, and Yule (1995) state that extending previous work on social support to the study of social support following a traumatic event is difficult, as traumatic events often interfere with the normal functioning of social support networks through the death of others or through disruptions caused by response to the incident. For example, McFarlane (1986) showed that families who had been exposed to the Australian bushfires were characterized by increased levels of irritability, fighting, withdrawal, and decreased enjoyment from shared activities.

In addition, quite a significant amount of literature has implied that social support does not need to be activated to be beneficial. Just knowing that others care for you during times of crisis reduces stress. In a study with married respondents in the United States, Wethington and Kessler (1986) were able to show that the stress-buffering effect of social support is more strongly linked to the perception that support is available than to the effects of actual supportive behaviors. Moreover, in some circumstances, the perception of social support that remains unutilized is more helpful than the actual receipt of social support. Several explanations for these findings are given, For example, overly intrusive social support can cause stress. The recipient may perceive efforts to provide support as controlling and interfering. The support that a social network member provides might be different from that which is needed (Taylor, Sherman, Kim, Jarcho, Takagi, & Dunagan, 2004). According to Bolger, Zuckerman, and Kessler (2000), awareness of receiving support entails an emotional cost, and the most effective support is that which goes unnoticed by the recipient. Otherwise, expressing one's need to others is esteem-reducing and/or drawing on another person for support is perceived to tax that other person's resources. These findings suggest that perceived social support is a stronger predictor of adjustment to stressful life events than received social support (Bolger et al., 2000; Taylor et al., 2004). Moreover, a study with Hispanic elders from East Little Havana, Florida, could demonstrate that satisfaction with received social support is associated with lower psychological distress, whereas the actual level of received social support is associated with heightened psychological distress (Cruza-Guet et al., 2008).

Pro Social Support

Even though there have been inconsistencies in research outcomes regarding social support, it is still considered an important predictor for stress-related disorders. For example, of 14 separate risk factors for PTSD investigated in a meta-analysis, social support was shown to have the strongest weighted average effect size (0.40) (Brewin et al., 2000). Furthermore, in the Detroit study by Breslau et al. (1998), a lack of social support has been proven to be a serious risk factor. If social support is perceived as missing, this poses a greater risk of individuals developing PTSD than the positive effect of social support being perceived as existing. This stronger impact of negative over positive support was confirmed by Andrews et al. (2003) in a study with victims of violent crime.

Subjectively perceived social support serves as an important protective factor against the onset of PTSD symptoms (Maercker, 1998). In a study with Israeli soldiers, Solomon et al. (1988) found that the greater the perceived social support, the less intense the PTSD three years after the war. A recent study with American war veterans who served in operations enduring freedom and Iraqi freedom led to similar results, as post-deployment social support predicted lower levels of traumatic stress (beta = -.31) and depressive symptoms (beta = -.21), even after controlling for demographic factors and combat exposure. Here, social support was defined as post-deployment emotional support, as well as instrumental assistance provided by family, friends, coworkers, employers, and community (Pietrzak, Johnson, Goldstein, Malley, & Southwick, 2009). Global perceived social support was strongest in predicting PTSD among women who reported both child abuse and adult rape. However, this buffering effect was not seen in non-abused women or women who suffered from either childhood or adulthood abuse. This finding implies the cumulative effect of perceived social support (Schumm, Briggs-Phillips, & Hobfoll, 2006).

Contra Social Support

However, a study investigating the psychological effects of the attack on the World Trade Center on September 11, 2001, showed that a low level of perceived social support only predicts major depression, not PTSD. A low level of social support might also be caused through PTSD or depression. Eventually, family members are overwhelmed with the symptoms and needs of the survivor and withdraw their help and support (Resick, 2003).

In a study with parents of chronically ill children in Iceland, the hypothesis that more perceived social support would predict less parental stress was not supported. Even though a lack of social support at the time of diagnosis predicted a greater impact of the disease on the family, it did not predict more trauma symptoms. However, the authors suggest that individuals exposed to more disease-related stressors might receive more social support and thus rate the impact of the disease as lower, even in the presence of stressors (Guðmundsdóttir, Elklit & Guðmundsdóttir, 2006). In a recent longitudinal study with male veterans conducted by Laffaye, Cavella, Drescher, and Rosen (2008), the hypothesis that social factors would influence the course of chronic PTSD symptoms was not supported. Levels of interpersonal resources and stressors at Time 1 did not predict Time 2 PTSD symptom severity. In Araya et al.' (2007) study with post-conflict displaced Ethiopians, the hypothesis that perceived social support would act as a mediator for the effects of trauma on mental distress was also not supported. Rather, perceived social support was associated with a higher quality of life. Another earlier longitudinal study by Murphy (1987) examined the role of self-efficacy and social support in mediating the effects of disaster stress on health one and three years after the volcanic eruption of Mt. St. Helens in the United States. The results indicated that self-efficacy was a significant predictor of mental distress, whereas social support was not. Social network size also more strongly associated with health outcomes than network contacts. It seems that if individuals believe there is someone to whom they can turn, this perception of potential support apparently affects the appraisal of the stressor as manageable (Murphy, 1987).

4.3.2 Social Support and the Family

It was also shown that strong kin relationships before the disaster are highly supportive and facilitate post-disaster recovery. However, if the kin live close by and are also affected by the disaster, they become less helpful (Smith, 1983). A kin-oriented rural community will be less well prepared, less flexible, and less quickly adjusted to the immediate aftermath, but will provide better long-term support through kin group interaction compared to an individuated urban, industrialized community (Hill and Hansen, 1962).

During the impact of a natural disaster, the response is aimed at personal survival. However, once the intense trauma of the impact is over, victims soon become concerned about family members, neighbors, and the community. The family is the major focal point and main source of help to victims during this period. Nevertheless, with sudden, widely devastating impacts, victims will be collectively and temporarily stunned for a brief period (Smith, 1983).

However, disasters such as the Indian Ocean tsunami illustrate quite clearly that the devastation is inflicted not only on individuals and families, but also on entire communities, and that the recovery process involves every social class, ethnic group, religious faith, and sector of society (Reyes, 2006).

Social networks and community resources, which usually help people to overcome the trauma, are destroyed. This leads to chronic psychosocial stress. Disaster victims also perceive the received social support as less (Herzog, 2004).

4.3.3 Community Support and PTSD

The surviving members of a community can form strong bonds and help each other. In 1985, Green, Wilson and Lindy mentioned the so-called *recovery environment,* which refers to the environment, in which the recovery after a trauma takes place. People from cultures with familial stability, a secure social environment, community support and low consumption of intoxicants are less vulnerable to PTSD, according to Fairbank, Schlenger, Saigh and Davidson (1995). Benight (2004) shows the effects of social support and other collective processes, for example, the efficacy of collective crisis management even under circumstances of high material loss. People have always sought close emotional relationships with one another in times of crisis. In this way, traumatic experiences can be integrated. Emotional attachment is considered an effective protective factor against helplessness and meaninglessness, according to McFarlane and van der Kolk (1996).

The role of social support was also highlighted in a study conducted by Armenian et al. (2002) assessing the occurrence of major depression after the 1988 earthquake in Armenia. Being with someone during the disaster, receiving assistance and support after the earthquake, and alcohol use were all protective against depression. As in the later study on the rate of depression, the intensity of loss was also found to be a risk factor for PTSD in an earlier study conducted by Armenian et al. (2000), and early support to survivors reduced the risk of PTSD. Those people who were accompanied by another person at the time of impact and who made new friends after the earthquake had the lowest PTSD rates.

4.3.4 Cultural Aspects of Social Support

Sociohistorical factors and cultural characteristics shape the giving and receiving of social support. The shared belief system of the group, along with attitudes, norms, and expectations, provides direction and guidance to the form in which social support is given and received (Dilworth-Anderson & Marshall,

1996). Moreover, research has suggested reliable cultural differences in how people view the self and their relationships that may have implications for whether or not they use social support to cope with stress. People in Western cultures tend to view themselves as independent and separate from other people, whereas Asians tend to view a person as fundamentally connected with others (Markus & Kitayama, 1991). This difference might lead to the assumption that coping via social support is especially common among Asians because they emphasize interconnectedness with their social group.

However, the opposite might be the case, according to Taylor et al. (2004). The idea that social support involves specific transactions whereby one individual enlists the help of another in service of his or her problems may be a particular Western conceptualization of social support. The independent view of the self holds that individuals take actions that are oriented toward the expression of their opinions and beliefs, the realization of their rights, and the achievement of their goals (Markus & Kitayama, 1991). According to this viewpoint, it is appropriate to enlist others in meeting those goals. Therefore, stressed individuals may focus primarily on themselves and their goal of coping with the stress, and recruit the time and attention of others in the process. In contrast, Asians tend to view a person as primarily a relational entity, interdependent with others. In these cultural contexts, social relationships, roles, norms, and group solidarity are typically more fundamental to social behavior than an individual's needs. This view of the self holds that a person should conform to social norms and respond to group goals by seeking consensus and compromise; as such, personal beliefs and needs are secondary to social norms and relationships. In Asian cultural contexts, given that emphasis is placed on maintaining harmony within the social group, any effort to bring personal problems to the attention of others or enlist their help may risk undermining harmony and/or making inappropriate demands on the group (Taylor et al., 2004). Similarly, in the study by Cruza-Guet et al. (2008) with Hispanic elders who originate from a more collectivistic culture, those persons who provided more social support than they received and those who engaged in equitable exchanges of social support reported lower levels of psychological distress. Chen and Mak (2008) investigated how cultural beliefs influence help-seeking behavior in a study with four groups of college students with varying degrees of acculturation to Western culture: European American, Chinese American, Hong Kong, and mainland Chinese college students. It was shown that the European American and Chinese American students, who were more influenced by a Western, individualistic culture, were more likely to seek help from mental health professionals than the Hong Kong and mainland Chinese students.

5 Etiology of PTSD

The question has arisen, to what extent personality and other predisposing factors play a role in the etiology of PTSD. When PTSD was included in DSM-III, there was a consensus that the nature and intensity of the traumatic stressor was the primary etiologic factor determining the symptoms. The same holds true for the ICD-10, where "a stressful event or situation (either short- or long-lasting) of an exceptionally threatening or catastrophic nature, which is likely to cause pervasive distress in almost everyone" (World Health Organization, 1992, p.147) leads to post-traumatic stress symptoms.

The previous classification systems tended to assume that symptoms would be time-limited unless some pre-existing character pathology was present that would contribute to their maintenance. This implied that the stress response was transient in nature unless the individual had some underlying vulnerability (McFarlane, 1990).

Originally, PTSD was considered a normal reaction to an abnormal event. However, not everyone develops PTSD symptoms after such an event. Clinicians have been uncertain about the normal reaction to a traumatic event and whether PTSD was a pathological state or not. Obviously, there will be a blurred division between those individuals who experience the normal stress response and those who have developed PTSD, particularly in the weeks following the traumatic event. This confusion is also embodied in the DSM-III-R (American Psychiatric Association 1987) criteria for PTSD, where PTSD is a legitimate diagnosis only if an individual has experienced an event that would be "markedly distressing to almost everyone" (p.250). In the DSM-IV, the subjective experience of the event is given more importance and the diagnose of PTSD is only given if "a person experienced, witnessed, or was confronted with an event or events that involved actual or threatened death or serious injury, or a threat to the integrity of self or others, and the person's response involved intense fear, helplessness or horror (in children, disorganized or agitated behavior)" (American Psychiatric Association, 1994).

That is why the traumatic situation is considered a trigger for a reaction that is influenced to a great extent by peri- and post-traumatic factors (Yehuda, 2000). Each person has his or her own personal history, individual experiences, coping strategies, and value system. As a result, the meaning that is given to a certain traumatic experience and the adaptation to this experience differs for each person (Krüsmann & Müller-Cyran, 2005).

Even though there are psychological responses as a direct consequence of a traumatic event, no linear relationship exists between trauma and response. According to Kudler (2000), there is no one right paradigm for PTSD, so it is necessary to flexibly shift between paradigms. Diverse approaches need to be accommodated and previous views should not be simplistically discarded as irrelevant.

5.1 Neurobiological Models

In order to understnad the etiology of PTSD symptoms, it is important to consider neurobiological factors that may be universally valid across cultures. This chapter shall give an overview on changes in the neurotransmitter system following the experience of a traumatic event. Furthermore, early biological predictors of PTSD may exist, for example, increased heart rate or lower urinary or plasma cortisol levels.

5.1.1 Neurotransmitters

Traumatic experiences overwhelm the individual's coping mechanisms and pose a threat to his or her very survival. Therefore, it is not surprising that the sympathic nervous system springs into action (Krüsemann & Müller-Cyran, 2005). Centers in the brain that deal with stress are activated, such as the locus coeruleus, hippocampus, amygdala, and the medial prefrontal cortex. A certain emergency mechanism begins to do its work, starting with the amygdala (Birbaumer & Schmidt, 2006). The amygdala is mainly responsible for the occurrence and regulation of emotions and the assessment of threat-related stimuli, and it is necessary for the process of fear conditioning. As a response to a stressor, it activates the locus coeruleus. This center provides the brain with noradrenalin, making a person alert and ready for fight or flight impulses.

A few minutes later, another reaction chain is triggered in the other primary stress pathway, the hypothalamic-pituitary-adrenal (HPA) axis stress-response cascade. The hypothalamus produces more corticotrophin-releasing factor (CRF), which promotes the secretion of adrenocorticotropic hormone (ACTH) from the pituitary gland. ACTH is transported through the blood stream to the adrenal gland, which then produces more cortisol. This reaction chain leading to increased arousal operates with every stress reaction. Normally these stress symptoms are reduced when the hippocampus and medial prefrontal cortex influence the amygdala to inhibit the release of CRF and ACTH (Roth & Münte, 2003). The activities of the amygdala are increased during traumatic stress, and the regulating activities of the hippocampus and cortex are reduced. In other words, the neurocircuitry model of PTSD postulates that the amygdala is hyperresponsive, the medial prefrontal cortex is hyporesponsive, and the medial prefrontal cortex and hippocampus fail to inhibit the amygdala. In addition, neuroimaging research provides evidence for diminished hippocampal volume and function (Pitman, Shin & Rauch, 2001). Animal studies have shown that high levels of stress-induced glucocorticoides in the blood stream have a toxic effect on the hippocampus (Sapolsky, Uno, Robert & Finch, 1990).

When the amygdala constantly fires warning signals, the individual experiences restlessness, the surroundings seem estranged, time seems to be delayed or suspended, and the event is beyond words or order. This way of experiencing when only the amygdala system and not the hippocampus or neocortex is reacting is called *peritraumatic dissociation* (Kapfhammer, Dobmeier, Ehrentraut & Rothenhäusler, 2001). This dissociation can also be caused by the release of the body's own opiates as a protective shield against overwhelming emotions and information. These opiates lead to psychomotor freezing, numbing, and a decreased feeling of pain and panic (Scaer, 2001).

Originally, PTSD was thought to be a natural process of adaptation to an unusually severe stressor. Moreover, many biological findings in PTSD are compatible with those of the chronic stress response, such as increased CRF concentrations, catecholamine depletion within the central nervous system, and reduced hippocampal volume. Chronic stress is expected to result in sustained cortisol increases. However, chronic PTSD was discovered to be distinctly different from other chronic stress responses, especially involving the HPA axis: urinary and plasma cortisol levels are considerably lower in PTSD patients than in non-PTSD trauma survivors and normal controls. This is due to a highly sensitized HPA axis characterized by decreased basal cortisol levels and enhanced negative feedback regulations with glucocorticoid receptors in target tissues that are highly sensitive to cortisol (Yehuda, 2001).

Acute biological responses after exposure may serve as risk or resilience factors for the development of PTSD. Hence, there are early biological predictors of PTSD, for example, increased heart rate or lower urinary or plasma cortisol levels. Moreover, prior trauma contributes to hormonal abnormalities (lower plasma cortisol levels) as well as increased risk for the development of PTSD following a subsequent trauma. Animal and human research has suggested that low cortisol levels may fail to keep sympathic responses in check, which leads to prolonged availability of noradrenalin in the brain. Consequently, traumatic memories are overconsolidated, which results in intrusions and avoidance behavior. However, a causal relationship between low cortisol levels and the onset of PTSD symptoms has not been determined (Delahanty & Nugent, 2006).

It remains to be clarified whether these effects in the brain are consequences of exposure to traumatic events, consequences of intrusive and hyperarousal symptoms, or somehow reflect pre-existing vulnerabilities that increase the probability of developing PTSD after a traumatic event (Ozer et al., 2003).

5.1.2 Memory

The majority of people who have experienced a traumatic event have the ability to somehow continue with their life and are not persecuted by memories of the event. Occupying oneself with the event and having intrusive memories may even be considered a normal way of coping with the trauma. If the emotions associated with the event are repeatedly aroused by memories of the incident, they can be altered and a tolerance for them is created (Horowitz, 1978).

However, according to van der Kolk (1996), when people concentrate selectively on reminders of the traumatic event and thus fixate on traumatic memories, new incoming information cannot be adequately processed. "Despite the human capacity to survive and adapt, traumatic experiences can alter people's psychological, biological, and social equilibrium to such a degree that the memory of one particular event comes to taint all other experiences, spoiling appreciation of the present" (van der Kolk, 1996, p.4). In the long run, the traumatic experience cannot be integrated and accepted as a part of one's past, and it starts to take on a life of its own, independent of previous cognitive schemata, in other words, it is dissociated. Specific patterns of avoidance and hyperarousal are then developed, and people start organizing their entire life around the trauma. Moreover, intrusive memories evoke intense emotions and sensations as if the event was happening all over again. The passage of time does not alter these memories in PTSD patients. In PTSD, the memories are not transformed (van der Kolk, 1996).

It is not easy to investigate trauma-related memories, as researchers cannot simulate a trauma in the laboratory. However, researchers can collect retrospective reports of traumatized individuals, conduct post-hoc observations, or provoke trauma-related memories and flashbacks in the laboratory (van der Kolk, 2000).

Lesions of the hippocampus and the medial prefrontal cortex lead to a strong impairment of the declarative or explicit memory system (van der Kolk, 1996). This memory system is connected to the conscious perception of facts or events that the individual has experienced. The implicit memory system, however, is connected to the memory of abilities and habits, emotional reactions, reflexes, and conditioned responses (van der Kolk, 2000).

In a traumatic learning situation, conscious memories are formed by a system involving the hippocampus and related cortical areas, and unconscious memories established by fear conditioning mechanisms operating through the amygdala-based system. These two systems store different kinds of information regarding the event and work parallel. When the individual encounters a stimulus representing the traumatic situation, the amygdala-based system causes him or her to respond physically (e.g. heart palpitations) and prepare for danger, whereas the hippocampus system leads to conscious memories (LeDoux, 1998).

5.2 Behavioral and Cognitive Models

In the classical behavioral model, which was used to conduct research on and treat Vietnam War veterans in the 1970s, the PTSD symptoms were believed to be caused by classic and operant conditioning. The traumatic event is considered the unconditioned stimulus causing immense fear, the unconditioned response. The fear of certain neutral stimuli associated with the traumatic situation is the conditioned response. The avoidance of these stimuli lessens the fear and thus serves as a negative reinforcer, according to the theory of operant conditioning. In this way, the fear of certain situations or thoughts is not erased.

Behavioral models of PTSD explain why people affected by PTSD suffer from anxiety and avoidance symptoms, but not why they suffer from intrusive memories (Resick, 2003). Cognitive theories, however, aim to explain such intrusive symptoms. According to Ehlers and Clark (2000), PTSD is a disorder that is associated with memories of a past event, as opposed to being anxious about a future event. In order to explain the development of chronic PTSD, the anxiety that PTSD patients experience must come from perceiving an immediate threat. Two processes cause this perception of an immediate threat: (1) excessively negative appraisals of the trauma and/or its consequences; and (2) a disturbance of autobiographical memory characterized by poor elaboration and contextualization, strong associative memory, and strong perceptual priming. A change in the negative appraisals and the trauma memory is prevented by a series of problematic behavioral and cognitive strategies. This model also provides a framework for cognitive-behavioral treatment of PTSD by identifying key targets for change.

5.2.1 Appraisals of the Trauma and/or its Consequences

People with chronic PTSD are not capable of viewing the traumatic experience as a time-limited event. Rather, they believe their entire life to be negatively affected by the event. Interpretations of the event and/or its consequences can be external (e.g. “There is danger everywhere” or “The next disaster will happen soon") or internal (e.g. "I'll never be able to lead a normal life again" or "I'll loose my job"). They lead to the feeling of a current threat. The individual's behavior during the traumatic event and the PTSD symptoms themselves can also be negatively interpreted. If individuals do not perceive these symptoms as a normal part of the recovery process, they may interpret them as indications that they have permanently changed for the worse or as indicators of a threat to their physical or mental health. Such appraisals maintain PTSD by directly producing negative emotions such as anxiety, depression or anger (Ehlers & Clark, 2000). Ehlers and Steil (1995) suggest that the idiosyncratic meaning of the intrusive recollections predicts the distress they cause and also the degree to which the individual makes an effort to control these intrusions. However, excessive control of the intrusive recollections will prevent a complete processing of the trauma and thus prolong the adaptation process. Therefore, the trauma will remain in active memory and intrusions will continue to occur. These control strategies also maintain the intrusive memories by preventing a change in the negative meaning of the trauma and traumatic memories.

5.2.2 The Characteristics of the Trauma Memory

PTSD patients often have difficulties intentionally remembering the complete traumatic situation. Their memories are often fragmented, details are missing, and the events are out of order. However, lively recollections of the trauma can come to mind against the individual's desire. These often take the form of sensory impressions such as visual images, sounds, or physical sensations. The individual experiences these memories as happening right now in the moment and experiences the same emotions as during the trauma. The emotions and physical sensations experienced during the trauma can also occur "out of the blue" without any conscious memory of the event. Ehlers and Clark (2000) suggest three ways of explaining the phenomena of these intrusive memories: (a) Traumatic memories are not processed elaborately enough and are not embedded into the structure of the autobiographical memory system. In other words, the meaning of the event is neither processed nor coded as having happened in a certain time period, nor linked to new information. Consequently, the memories can be easily triggered; (b) especially strong stimulus-stimulus and stimulus-reaction associations are formed as part of the PTSD. Certain stimuli perceived right before or during the traumatic situation are associated with a severe threat to the self. When confronted with such a stimulus, the individual has the feeling of reliving the trauma; (c) strong priming has taken place in the implicit memory system for stimuli associated with the trauma. Patterns of stimuli that have been perceived before have a lower threshold of being perceived as new stimuli. For this reason, stimuli that can trigger the reliving of the traumatic situation are easily noticed, even if they are only vaguely similar to the original stimuli.

5.2.3 Dysfunctional Behavior and Cognitive Processing Styles

Certain cognitive strategies to diminish the perceived threat are called dysfunctional because they enhance the PTSD symptoms. For example, people who believe that intrusive memories mean they are losing control of their mind are likely to atttempt to erase such recollections from their mind. Unfortunately, active thought suppression leads to the thoughts becoming more frequent. Other behavior that is supposed to reduce specific PTSD symptoms can increase other symptoms. For example, going to bed late to avoid nightmares can lead to irritability, lack of concentration, or loss of interest. Avoidance of thoughts about the trauma prevents the elaboration of the trauma memories; it also prevents the falsifying of negative interpretations. Furthermore, if stimuli and places associated with the trauma are avoided, the interpretations cannot be modified. Ruminative thoughts such as "What if..." are also not helpful, as they lead to feelings of helplessness, depression, anger, or nervous tension. Other dysfunctional behavior is alcohol or medication abuse, or the avoidance of previously cherished activities (Ehlers & Clark, 2000). The following coping strategies are especially likely to lead to the maintenance of intrusive symptoms: safety behaviors, alcohol and drug (ab)use, dissociation and numbing, thought suppression, rumination, activation of other thoughts such as anger and guilt and corresponding cognitions, and selective information processing (attention and memory biases) (Ehlers & Steil, 1995).

Dunmore et al. (2001) investigated several cognitive factors highlighted in Ehlers and Clark's (2000) cognitive model of persistent PTSD using a prospective design. Cognitive factors of 57 victims of physical or sexual assault were assessed within four months of the assault and followed up after six and nine months. Several cognitive factors were found to predict PTSD at both follow-ups: cognitive processing style during assault (mental defeat, mental confusion, and detachment); appraisal of assault sequelae (negative appraisal of initial PTSD symptoms, negative perception of other's responses, and perceived permanent change); negative beliefs about self and the world; and maladaptive control strategies (avoidance and safety seeking). Even after controlling for gender and perceived assault severity, the relationship between early appraisals, control strategies, processing styles and subsequent PTSD levels remained significant, thus supporting the cognitive model.

In Ehlers et al.'s (1998) prospective longitudinal study with 967 patients who attended an emergency clinic after a motor vehicle accident, the most important predictors for the presence of PTSD symptoms at one year after the accident were found to be a negative interpretation of intrusions, persistent medical problems, and rumination at three months.

Research has been conducted to investigate how thought suppression leads to more frequent intrusions. In a similar manner, attempts to suppress traumatic material may be involved in the development and maintenance of PTSD symptoms. Davies and Clark (1998) induced analogue post-traumatic intrusions in normal participants by showing a distressing film in which people are shown dying in a building fire. A film about polar bears was also shown. The participants were randomly assigned to two groups: the first group was instructed to try to suppress thoughts about the distressing film, but only during the first thought-recording period, and the second group received no such instructions. Both groups were instructed to record all thoughts during the two time periods. The results indicated that the thought suppression group was initially able to suppress thoughts about the distressing film.However, during the second recording period, they reported more thoughts about the distressing film than the control group ("rebound effect"). No such effects were found for the polar bear film.

6 Research Questions and Hypotheses

Already existing studies investigated the consequences of comparable flood-disasters in different countries. The results were that flood victims showed more depressed symptoms, anxiety, traumatic stress, and emotional disorders than non-flood victims (McMillen, North, Mosley, & Smith, 2002). However, it has yet to be determined to what extent the indigenous peoples in tsunami-affected areas are still traumatized two and a half years after the disaster. In this context, the cross-cultural validity of the PTSD concept needs to be further explored. In refugee groups from non-Western countries, for example, using the PTSD-checklist, there was a high construct-validity of the items in the symptom group "intrusion" and "increased arousal" However, this was not the case with the symptom group "avoidance/emotional numbing" (Joannidis, 2006). These ethno-cultural aspects have so far been neglected. Moreover, the pathogenetic processes leading to PTSD symptoms are well investigated, but not the personal resources such as a SOC, certain coping strategies, and perceived social support that act as protective factors, especially in the case of flood disaster victims (Otto, Boos, Dalbert, Schöps & Hoyer, 2006). Furthermore, the findings published on how variables such as the severity of trauma (e.g. whether participants lost a family member, their home, or their means of earning a livelihood) or demographic factors (e.g. level of education, age, or marital status) influence the degree of PTSD are far from being consistent. In order to contribute to clarifying these issues, the following research questions have been derived. Moreover, for each of these questions a summary of the relevant and controversial findings will be given.

6.1 Research Questions

1. Do tsunami victims show higher rates of PTSD than non-tsunami victims two and a half years after the tsunami?

Depending on the kind of disaster or study, between 5 and 60 percent of the victims develop PTSD or other psychiatric disorders. According to Kessler et al. (1995), the probability of developing PTSD after a natural disaster or fire is 5.4 percent for women and 3.7 percent for men (Kessler et al., 1995). In a sample involving child and adolescent survivors of a super-cyclone in Orissa, India, 30.6 percent were diagnosed with PTSD according to the ICD-10 criteria of the WHO (1992) and an additional 13.6 percent from sub-syndromal PTSD (Kar et al., 2007). Wang et al. (2000) also investigated the levels of earthquake-related PTSD in a community sample in North China. The rate of onset of DSM-IV PTSD symptoms within nine months of the earthquake was 19.8 percent and

30.3 percent, respectively, in two villages, one mildly affected and one severely affected. The DSM-IV criterion requiring clinically significant distress or impairment in functioning for a diagnosis of PTSD was a major contributor to the lower rates of diagnosed PTSD. Otherwise, the pooled rate of PTSD in the two villages would have been 35.0 percent. According to DSM-III-R criteria, the pooled rate of PTSD would have been 41.4 percent. In 1987, Madakasira and O'Brien (1987) found that 59 percent of the interviewed survivors of a tornado in the United States qualified for a diagnosis of PTSD according to DSM-III criteria, of whom 16 percent had a severe form.

In another U.S. study, Green et al. (1992) found that 20 years after a devastating flood, twenty-five percent of the affected people still suffered from PTSD. Often, the PTSD symptoms disappear after a few weeks, but if they remain for more than three months, the prognosis is unfavorable, as there is an increasing danger of the symptoms becoming chronic (Frommberger & Maercker, 2006). Moreover, the National Comorbidity Survey by Kessler et al. (1995) showed that in 30 percent of the cases the symptoms decreased after 12 months, in 20 percent of the cases after four years and in more than one-third of the cases the symptoms persisted for more than six years. According to Breslau et al. (1998), in more than one-third of the cases the symptoms persisted for over five years.

2. *Do tsunami victims show higher rates of general psychological distress than non-tsunami victims two and a half years after the tsunami?*

Various studies demonstrated that PTSD often accompanies other disorders such as anxiety and affective disorders, substance abuse/dependence, somatization, as well as antisocial personality disorder (Breslau et al., 1997; Kessler et al. 1995). Kessler et al. (1995) found that 79 percent of the women and 88.3 percent of the men suffering from PTSD had at least one other mental disorder throughout their life. Green et al. (1992) interviewed 193 survivors of the breach in the Buffalo-Creek dyke in the United States. Fourteen years after the flood, twenty-five percent of the population still showed symptoms of PTSD. Forty-two percent of these PTSD patients developed comorbidity with depression, 42 percent with generalized anxiety disorder, and 29 percent with simple phobia.

A study by Kumar et al. (2007) with 314 victims of the 2004 tsunami in Tamil Nadu revealed a PTSD prevalence rate of 12.7 percent, measured with the Harvard Trauma Questionnaire two months after the disaster. Among individuals who sought help from psychiatrists (15% of the overall sample), 31.9 percent were diagnosed with PTSD and 48.9 percent with major depression. In a recent study with Sri Lankan tsunami victims conducted by Hollifield et al.

(2008), the prevalence of clinically significant PTSD, depression, and anxiety was 21, 16, and 30 percent respectively. The measures used were an adapted questionnaire assessing PTSD according to ICD-10 and the Hopkins Symptom Checklist-25 (HSCL-25). Moreover, participants reported a mean of eight somatic complaints associated with psychological distress

3. *What is the relationship between Antonovsky's Sense of Coherence (SOC) (Antonovsky, 1987) and the level of PTSD? Does the SOC function as a buffer?*

One protective factor that is supposed to act as a buffer to mitigate the effects of the traumatic event is the sense of coherence. The SOC refers to people's ability to mentally sort, understand, and give meaning to events (Antonovsky, 1987). This ability has a positive effect during and after the trauma. The empirical evidence suggests that the degree of SOC is negatively correlated with the degree of PTSD (Maercker, 1997).

A study conducted by Frommberger et al. (1999) showed that cognitive and emotional reactions to a motor vehicle accident strongly depended on the people's SOC, independent of the severity of the accident. Several studies demonstrated a positive correlation between the existence of a sense of coherence and psychophysical health and adequate coping (Höge, 2005; Groß & Pöhlmann, 2004; Schüssler, 2003; Sek & Tasikowski, 2002). According to the salutogenesis model, a person can be placed somewhere along a continuum between healthy and sick. In which direction he or she moves after an encounter with a trauma, depends on the balance between the severity of the stressor and the SOC (von der Heyden, 2006).

4. *What is the relationship between social support and the level of PTSD? Does social support act as a buffer to the severity of the trauma?*

Social support can influence mental and physical health in two ways: firstly, it can directly lead to greater well-being (main effect), and secondly, it can modify the effect of stressors on well-being (buffer effect). According to Lazarus and Folkman's (1984) transactional stress model, the buffer effect can be explained as follows: perceived social support has a positive effect on the primary and secondary assessment of a stressor, since people who have adequate social support believe that they have access to a larger number of coping strategies (Schützwohl, Maercker, & Manz, 1999).

A lack of social support has been proven to be a serious risk factor (Brewin et al., 2000), whereas subjective perceived social support serves as an important protective factor against the onset of PTSD symptoms. Here social support means the presence of emotional, practical and material support. The stronger the perceived social support, the lower the risk of developing PTSD (Maercker, 1998). However, a study investigating the psychological effects of the attack on the World Trade Center on September 11, 2001, showed that a low level of perceived social support only predicts major depression, not PTSD (Galea, Ahern, & Resnick et al., 2002). A low level of social support might also be caused by PTSD or depression. Eventually, family members are overwhelmed with the symptoms and needs of the survivor and withdraw their help and support (Resick, 2003).

5. *Which role do coping strategies play? Which coping strategies are more effective? Is avoidance coping more effective in reducing PTSD than approach coping, as this study was conducted in a collectivistic cultural setting?*

According to the transactional model of stress and coping (Lazarus & Folkman, 1984), coping strategies are activated when internal or external stressors exhaust the resources of an individual. The different coping strategies can be divided into two main groups: problem-focused coping strategies aimed at concretely solving the problem, and emotion-focused strategies aimed at reducing tension by denial or changing one's attitude (Folkman & Lazarus, 1980, 1985). In addition to the problem-focused versus emotion-focused coping approach, the other main classification of coping behavior is in approach versus avoidance coping. Several studies have shown that people who tend to engage more in approach coping, and less in avoidance coping show lower levels of PTSD symptomology. An avoidance coping style is generally considered a negative predictor of PTSD, for example, in victims of domestic violence (Krause et al., 2008) or in victims of Hurricane Katrina (Glass et al., 2009). Avoiding feelings, thoughts and actions associated with the traumatic situation does not lead to successful coping with the trauma (Dunmore et al., 2001; Solomon et al., 1988).

However, palliative or avoidance strategies may be more effective in collectivistic cultural contexts where the norm is to "fit in" with the social and physical environment. Thus, coping strategies focusing on controlling one's own internal states and behaviors are deemed more desirable and may be more effective in achieving coping goals. For example, Chang (2001) investigated the normative coping strategies of Euro-American and Asian American college students and their relationships to positive and negative psychological outcomes. Compared to Euro-American students, Asian American students reported a greater use of problem avoidance and social withdrawal. The use of these avoidance strategies was associated with less life satisfaction and more depression in Euro-Americans, but it was not associated with either of these outcomes in Asian Americans. Likewise, Yoshihama (2002) found that the perceived effectiveness of passive coping strategies was associated with lower psychological distress in the more collectivistic Japanese-American women born in Japan, whereas the perceived effectiveness of active coping strategies was associated with lower distress in their more individualistic counterparts born in the United States. Furthermore, active coping appeared to have deleterious effects on the psychological well-being of the Japan-born women.

6. *Is coping triggered by the traumatic event, so that more coping responses take place after the tsunami? Does the onset of PTSD symptoms trigger more coping efforts?*

The empirical evidence suggests that the extent of coping efforts correlates positively with the degree of PTSD. For example, in a study with World War II veterans, Fairbank et al. (1995) showed that former prisoners of war diagnosed with PTSD used a broader spectrum of coping strategies and used these more often than prisoners of war without PTSD.

In a longitudinal study with American Vietnam War veterans suffering from PTSD symptoms Tiet et al. (2006) found that a high level of PTSD symptoms predicted more approach coping and more behavioral avoidance coping, which may be due to the fact that patients with more PTSD symptoms are expected to experience more stressors, which elicit more coping responses.

7. *What is the relationship between SOC and the coping strategies?*

The cognitive models of stress (Lazarus, 1999) focus on cognitive appraisals and coping, with personal resources (protective factors) being the antecedents that affect people's appraisals of their ability to cope with stressful situations. In this context, SOC is considered a personal resource (Moos & Schaefer, 1993) that increases the effectiveness of an individual's reactions to a stressful event. Antonovsky (1987, 1993) developed the SOC to expand the understanding of individual variability in reaction to stressful life events. According to this model, SOC is a personal resource that facilitates efficient coping with negative life events and stressors, buffering their effect on psychological well-being. When facing stressful situations, individuals with higher SOC reported having better coping resources and expressed more confidence in their ability to cope. Moreover, the studies also reported that individuals with higher SOC used more active and problem-focused coping strategies than individuals with lower SOC, who used more avoidant and emotion-focused coping. (Amirkhan & Greaves, 2003; Pallant & Lae, 2002). In an Israeli study, Cohen, Ben-Zur, and Rosenfeld (2008) showed how coping strategies mediated the effects of SOC on test anxiety and test performance.

A SOC is developed throughout childhood and young adulthood on the basis of life experiences. According to Antonovsky (1987), generalized resistance resources, such as social support, cultural stability, religion, and financial security, contribute to its development. Additionally, higher SOC was found to be related to demographic characteristics such as male gender or higher education (Suominen, Blomberg, Helenius, & Koskenvuo, 1999).

8. *Which demographic factors serve as risk factors, which as protective factors?*

Demographic and personality factors increase the probability of suffering from PTSD, such as low intelligence or a lack of education, being female, a pre-existing psychiatric illness or being young (meta-analysis by Brewin et al., 2000), or old age (Maercker, 1998). Widowed participants also d a significant higher score on depression as well as anxiety measures in the study with tsunami victims in Kerala conducted by Sunidharan (2005). In the study with Tamil tsunami victims by Kumar et al. (2007), the odds ratio analyses revealed that persons without income, women, and those who were injured during the tsunami were almost three times as likely to develop PTSD symptoms.

9. *To what extent does the severity of the trauma (e.g. loss of house or family members) affect the degree of PTSD?*

In Russoniello et al.'s (2002) study with children affected by Hurricane Floyd, children whose homes were flooded were three times as likely to develop symptoms of PTSD as children whose homes were not flooded. In the study by Kumar et al. (2007), those participants who had lost a family member were twice as likely to develop PTSD, and those who did not receive adequate support after the tsunami nearly one and a half times. In the study with survivors of a super-cyclone in Orissa, India, it was revealed that in higher exposure areas, more participants fulfilled the ICD-10 criteria for PTSD than in lower exposure areas (43.7% versus 11.2%) (Kar et al., 2007). Wang et al. (2000) investigated the levels of earthquake-related PTSD in a community sample in northern China. The results suggested that the village with a higher level of earthquake exposure and post-earthquake support had lower levels of PTSD than the village with a lower level of exposure but less post-earthquake support.

10. *What are the indirect and direct effects of social support and SOC on PTSD and other psychopathological symptoms? Is the effect of SOC on PTSD and other psychopathological symptoms both direct and mediated by coping behavior, i.e. approach and avoidance coping? Does the variable "being affected by the tsunami" have a direct effect on coping behavior and social support, as well as on both variables of traumatic*

stress? Does approach coping increase PTSD and other psychopathological symptoms? Does avoidance coping, in contrast, reduce variables of traumatic stress? What is the effect of social support on coping behavior?

These questions present the summary of the aforementioned ones and shall be explored by testing a path analysis model.

6.2 Hypotheses

1. Tsunami victims show higher rates of PTSD (assessed with the Impact of Event Scale-Revised (IES-R) by Weiss and Marmar (1996)) than non-tsunami victims two and a half years after the tsunami.
2. Tsunami victims show higher rates of general psychological distress (assessed with the Brief Symptom Inventory (BSI) by Derogatis (1993)) than non-tsunami victims two and a half years after the tsunami.
3. There is a negative correlation between Antonovsky's Sense of Coherence (SOC) (Antonovsky, 1987) and the level of PTSD.
4. There is a negative correlation between perceived social support and the degree of PTSD.
5. Avoidance coping is more effective in reducing PTSD than approach coping, as this study was conducted in a collectivistic cultural setting.
6. More coping efforts took place after the tsunami, i.e. the tsunami victims scored higher on all coping factors than the non-affected group.
7. SOC is positively correlated with more effective coping strategies and serves as a predictor for these coping strategies.

8. 1 The following demographic factors serve as protective factors, i.e. they are negatively correlated with PTSD and other psychological distress and predict a lower level of symptoms: high education, male gender, being married, middle age, or having received a new house and/or a new job

8. 2 The following demographic factors serve as risk factors; i.e. they are positively correlated with PTSD and other psychological distress and predict a higher level of symptoms: low education, female gender, widowhood, young or old age, or chronic illness.

9. Factors determining the severity of the trauma, such as loss of family members or friends, loss of home, damage of home, loss of job, or personal injury, predict a higher level of PTSD symptoms and other symptoms of psychological distress.
10. Social support has a direct effect on PTSD and other psychopathological symptoms. The effect of SOC on PTSD and other psychopathological symptoms is both directly and indirectly mediated by coping behavior, i.e. approach and avoidance coping. Whether people are affected by the tsunami or

not has a direct effect on coping behavior and perceived social support, as well as on both variables of traumatic stress. Approach coping positively predicts PTSD and other psychopathological symptoms, whereas avoidance coping negatively predicts variables of traumatic stress. Perceived social support has an effect on coping behavior.

7 Methods

In order to verify the first set of hypotheses (*Tsunami victims are psychologically more distressed in the long term than non-tsunami victims.)*, the means of the sum total of the test scores of the control group (no tsunami) and the treatment group (tsunami) are compared to determine whether there is a significant difference. Whether a village was affected by the tsunami or not is the independent variable. The dependent variables are the mean test scores of the questionnaires.

The second set of hypotheses (Protective factors such as a sense of coherence, perceived social support, and certain coping strategies decrease the extent of traumatization.) will be tested using Pearson's product moment correlations. This method will highlight the connections between psychological distress and health and protective factors. Finally, a multiple stepwise regression will show which variables contribute the most to the general variance.

In addition, the subjects were given a demographic sheet to assess the severity of their personal trauma, for example, whether they lost a family member, their home, or their means of earning a livelihood. The villagers were also questioned about their level of education, age, marital status and other demographic data. In this way, differences between control and treatment group could be ruled out and demographic data, as well as the severity of trauma, could be included in the regression analysis. This provided give insight into why some people are more affected by the trauma than others.

Finally, in order to investigate the last hypothesis and determine the complex relationships between several variables, a path analysis model will be tested.

7.1 Setting

In order to understand the impact of the tsunami in a particular region, it is necessary to consider the socio-cultural aspects of the setting. The study was conducted in Kerala, a state on the southwest coast of the Indian subcontinent. It is one of the densest populated states of India, with 844 people per square kilometer. In the local language, Malayalam, "kerala" means "land of the coconut trees," and coconut trees literally make up the largest part of the vegetation. Eighty-six percent of the land is used for agriculture. The main crops are rice, coconuts (45% of the harvest in India), cashew nuts, coffee, and cocoa. Industry only contributes to 10 percent of the annual income of the state. "Malayalam" means "land abounding in hills".

For hundreds of years, all of the major religions have peacefully coexisted. Of the population in Kerala, 56.2 percent are Hindus, 24.7 percent are Muslims, and 19.0 percent are Christians. The climate is tropical without much change in temperature throughout the year and is strongly influenced by the monsoon season, which starts beginning of July and can last until October. The literacy rate is the highest in India at 94.59% (men 94.2% and women 87.7%). Compared to other parts of India, there are no extreme cases of poverty in Kerala, for example, there are no slums. Furthermore, the educational and health care systems are more advanced in Kerala than in other Indian states. The life expectancy of 73.5 years (men: 70.6 and women: 76.1 years) is much higher than the average life expectancy in India of 61.7 years (men: 60.8 and women: 62.5 years). The ratio of women to men is 1,058 women to 1,000 men, whereas the average in India is 933 women to 1000 men (The Official Web Portal of Government of Kerala, 2009).

7.1.1 Social Structure in an Indian Village

In an Indian village, where the majority of its inhabitants are engaged in some agricultural activity and the people are highly dependent on each other, a strong sense of unity must exist. However, even today, caste is an institution of great strength, and as marriage and eating together are forbidden with members of other castes, the members of a caste have strong bonds with fellow caste members in neighboring villages. The caste ties are stronger than living in the same village. In this way, the villages are interdependent. A caste is not self-sufficient but depends on other castes that may have different occupations. Members of different castes can be linked in various ways, for example, the relationships between a landlord and tenant, master and servant, creditor and debtor. Any caste can be the dominant one (except the Untouchables) numerically and/or economically. The dominant caste supports and maintains the system and also resolves disputes among the other non-dominant castes. Before British rule, territorial divides (villages, chiefdoms, kingdoms) were more important than caste divides. However, with the British rule came a greater solidarity among castes. Thus, the British weakened vertical ties and strengthened horizontal ones. In this way, the organization of an Indian village reveals a more horizontal than vertical unity (Srinivas, 1969a).

In Kerala's dispersed villages, it is difficult to distinguish one village from another because of the scattered settlement. Economic unity may be modified by the extension of caste obligations to several villages. For example, the official astrologer family of the village may also serve other neighboring villages. Therefore, it is difficult nowadays to point to any unit as a clearly coherent, independent village community. The villager's social field is much wider than his village, as it is enlarged by kinship, economic, religious and other social ties to a circle of neighboring villages. Consequently, the local community does not necessarily coincide with any administrative unit. In some circumstances, however, the often invisible village unity can resurface.

> When the village is threatened with an epidemic or drought or flood or fire, or when the government passes an order which the villagers regard as unjust, or on certain religious occasions, or in a fight with a neighboring village, the unity of the village reveals itself in an unmistakable manner (Srinivas, 1969a, p. 8).

According to Srinivas (1969b, p.21), "A feature of the village community is the *panchāyat* or council of elders which decides disputes between villagers, and discusses matters of common interest such as holding a festival and building a temple or road". It is the smallest entity of India's political system of decentralization (Srinivas, 1969b; Pandey, 2006).

The joint family has long been the common form of family organization in India, sanctified in the scriptures and sanctioned in secular law. It consists of several nuclear families living together under one roof, usually the father and son or brothers with their wives and children. The property of everyone is held in common under the trusteeship of the oldest male family member. Every male child is entitled to a share of the joint family property. Every member of the joint family is fed from a single kitchen and receives money from the family purse. Among cultivators, everyone in the joint family work together for the family's crop (Mandelbaum, 1969). After marriage, the newly-weds live with the groom's family. The mother in the Indian family is given respect and almost a goddess-like status, at least theoretically, and she is responsible for taking care of the family members' need for love, care, and protection. However, she does not decide important matters of the household as this is the father's role. He is also the one who earns a livelihood. The grandparents traditionally have an important role in the family, as they are the custodians of culture and conventions. They take care of the grandchildren in the absence of the parents (Pandey, 2006).

7.1.2 Religion and Ideology in Kerala

Traditional religious festivals play an important role in daily life in Kerala. However, the purely religious meaning is gradually fading, and the festivals take on a more social meaning, for example, that of a unified state, as all people, regardless of their caste and religion, participate. In Kerala, diverse religions can be found and also a plurality within each religion can be found: Hinduism, Christianity, even within Catholicism (the various rites, Latin, Syro-Malabar, Syro-Malankar), Judaism, and Islam. Different castes and ethnic groups coexist, and for each group, religion is an important component of the group-specific culture.

One study was conducted in Kerala by Lemercinier (1983) to investigate the social functions of religion in the different modes of production. The author's hypothesis is that, in Kerala, religion performs the function of providing social identity. Religion forms a frame of reference within which the group understands its past, interprets its present and foresees its future, in terms of particular beliefs or forms of expression. By affecting social structures, religion provides the material basis for the identity of the group. However, in Kerala, no single religion or all-inclusive ideology was needed to integrate all of the social groups and reduce social oppositions.

The structure of the different types of social systems that followed one another during the period preceding the 19th century was also analyzed in Lemercinier's (1983) study: the system of kinship relations in the clans at the beginning of the Christian era, the political system of the monarchies in the first millennium, and finally, the religious system in the caste society which was built up gradually from the 12th century and continued to dominate until the 19th century. Apparently, the pre-existing social groups from the kinship mode of production were not destroyed throughout history. The objective of this study is to uncover the inter-relationships between the dominant system of social relations and the symbolic systems in order to determine if and how religion has an identity building function.

Collective interests are not completely identical with individual interests. In lineage-based societies, the survival of the group depends on the entire clan. For this reason, the clan - like any other social ensemble - requires a political authority and an ideology. For each individual, self-identity means viewing oneself as naturally included in one particular group and naturally excluded from other groups. The inclusion implies solidarity with the others in order to produce or reproduce the particular structure of the clan. Since the entire symbolic production is expressed in a religious code, identity must also be included in the same category.

7.1.3 Cultural Beliefs about the Tsunami

In India, the fisherfolk claim that the world is supported by a giant curled-up serpent. It is a symbol for a subtle energy that surrounds the world like a network. If the serpent's breath does not flow rhythmically, chaos is created in the world, and the Earth trembles and the sea spills onto the land. The serpent lies beneath the mountains, the fields and the sea deep inside the earth. Its rhythm is destroyed when the sensitive points along its back are covered or burdened. These points are knots of the net that encompass the Earth, and are guarded by Devas. However, the Devas cannot protect the serpent when hotels or factories are built on these points. For this reason, the villagers would never build houses in certain places. They also perform rituals and offerings to appease the serpent, so that it does not eat the fishers on the open sea (Link, 2005). Therefore, they see the reason for the tsunami as nature's wrath, caused by selfishness and greed of humans, who are destroying the Earth (Schnibben, 2005).

7.2 Procedure

In order to recruit the participants of the study, the author traveled to the villages along the coastline of the Alappad District, Kerala, and visited the families in their homes with a translator, who translated what she said in English into Malayalam, the language of Kerala. In addition, training was given to the translator so she could provide instructions on how to complete the psychometric scales. The translator was a woman from the village with whom the subjects were familiar, thus they were willing to participate in the study. In India, it is considered impolite to refuse a neighbor's request for help. Therefore, the subjects willingly completed the psychometric scales. Only one or two people refused to participate in the study. Not all questionnaires could be used for the data evaluation, however, as not all of them were complete. Of all the questionnaires distributed, approximately 64 percent could be used for the study.

All of the questionnaires were translated into Malayalam and then back-translated into English. These sets of questionnaires, which consisted of 12 pages each, were distributed in two villages affected by the tsunami (Azheekkal and Srayickadavu) and another village only indirectly affected by the tsunami (Pandarathurathu). In the villages affected by the tsunami (treatment group), the author selected the areas worst hit by the disaster. One hundred and nineteen villagers had received new houses, and 30 were still living in temporary shelters. The sampling method was cluster-sampling, as in these certain selected areas all adult inhabitants of the area were included in the study. The village only indirectly affected by the tsunami (control group) is located seven kilometers away from the tsunami-affected area along the coastline, but the water did not damage many houses there. The impact of the waves was broken by a dam, thus the water level did not exceed one meter in height. Obviously, the residents of Pandarathurathu were also affected by the tsunami, as the villages are in close proximity of each other and the fishing villages in Kerala are very interconnected. Many people lost relatives who lived in a neighboring village, which was affected by the tsunami or their family property that was located in another village was damaged. However, for the purpose of simplification, the author will refer to one group of participants as "affected by the tsunami" and the other group as "not affected by the tsunami." The villages of the control and treatment group are similar in terms of demographic factors such as education, mean age, and occupation, since all villages are located along the same coastal strip and the way of living (earning one's livelihood through fishing, everyday life) is virtually the same in the entire area. Given that the villagers were not very familiar with psychometric research methods, the translator explained to them the goal of the study and how to complete the questionnaires.

7.3 Participants

In order to determine the long-term consequences and protective factors after the tsunami in Kerala, India, the people directly affected by the disaster were asked to participate in the study along with a control group who were not affected. Subjects of the study were thus inhabitants of the fishing villages, Azheekkal, Srayickadu, and Pandarathurathu, along the coastline of Alappad District, Kerala.

Altogether, 407 subjects completed the questionnaires and thus could participate in the study. Two hundred and thirty-three people were affected by the tsunami (57.4%) and 174 were not affected (42.6%). Of these, 50.6 percent were male, and 49.4 percent were female. The mean age was 42.03 years, ranging from 15 to 90 years (SD = 14.45). Of the participants, 13.6 percent were single, 74.9 percent were married, and 11.5 percent were widowed. A total of 35.9 percent went to school up to the 8th standard, 46.3 percent up to the 10th standard, 11.4 percent completed the 12th standard, and 6.5 percent held a college degree. Of the men, 8 percent were unemployed, 75.1 percent were fishermen, 12.9 percent had some other occupation and 4 percent were still in school or college. Of the women, 86 percent were housewives, 12 percent had some other occupation, and 2 percent were students enrolled in school or college at the time of the data collection.

The following Table 2 gives an overview of the demographic data of the affected group, the not affected group, and the entire group.

Table 2

Overview of the Demographic Data for the Affected, not Affected, and Total Group

	Affected	Not affected	Total
	Mean (SD)	**Mean (SD)**	**Mean (SD)**
Age	43.6 (13.7)	40.0 (15.2)	42.0 (14.5)
	%	**%**	**%**
Gender			
Male	47.2%	52.6%	49.5%
Female	52.8%	47.4%	50.5%
Marital status			
Single	10.5%	17.6%	13.6%
Married	72.4%	78.2%	74.9%
Widowed	17.1%	4.2%	11.5%
Profession			
Housewife/unemployed	47.8%	45.7%	46.9%
Fisher	43.0%	30.6%	37.7%
Other job	5.7%	21.4%	12.5%
In school/college	3.5%	2.3%	3.0%
Level of education			
Up to 8th standard	45.5%	23.0%	35.9%
Up to 10th standard	45.5%	47.3%	46.3%
Up to plus two	7.2%	17.0%	11.4%
College	1.0%	5.4%	3.0%
Chronic illness			
Yes	75.0%	70.5%	73.1%
No	25.0%	29.5%	26.9%
Loss of family members or friends			
Yes	62.5%	16.8%	43.0%
No	37.5%	83.2%	57.0%
Loss of home			
Yes	75.4%	38.0%	59.6%
No	24.6%	62.0%	40.4%
Loss or damage of home	96.6%	87.2%	90.5%
Loss of job			
Yes	37.1%	11.0%	25.9%
No	62.9%	89.0%	74.1%
Personal injury/disabled			
Yes	68.9%	9.2%	43.1%
No	31.1%	90.8%	56.9%
New house			
Yes	53.2%	19.1%	38.7%
No	46.8%	80.9%	61.3%
New job			
New job	19.9%	2.9%	12.6%
No new job/same job	75.8%	97.1%	84.9%

There were significant differences regarding the demographic variables between the affected and not affected group, with the exception of gender and chronic illness (measured with a chi-square test). Differences as to age (measured with a t-test) and the damage of one's home (measured with a chi-square test) were significant with $p < 0.05$, and as to all of the other demographic variables with $p < 0.01$ (measured with chi-square tests). Nevertheless, the two groups are from a homogenous population of mostly fishermen and their families. Moreover, they live in the same coastal area and their daily lives do not differ. Therefore, the differences in demographic variables between the two groups are due to sampling methods and do not need to be given importance.

Table 3 gives an overview of the differences in occupation according to gender.

Table 3

Overview of Profession According to Gender

	Men	Women
Profession		
unemployed/housewife	8.0%	86.0%
fishermen	75.1%	0.0%
other profession	12.9%	12.0%
in school/college	4.0%	2.0%

In the affected group, 62.2 percent of the participants reported having lost family members or friends, 75.1 percent reported the complete loss of their home, 24.5 percent reported damage of their home, and 67.4 percent reported having been injred during the tsunami. Of these, 53.2 percent had received a new home and 19.9 percent a new job.

Before distributing the questionnaires, the author conducted a pilot study with 19 subjects from Parayakadavu, a village next to Srayickadu, which was moderately affected by the tsunami. The goal of the pilot study was to determine whether the participants would understand the concept of answering psychometric scales. The pilot study was successful, as the subjects had no difficulty in answering the items.

7.4 Instruments

Questionnaires can be a useful aid in gaining information about the symptoms, and it may be easier for some people to complete a questionnaire than to discuss their difficulties (Butollo, Krüsmann & Hagl, 1998).

In order to test whether tsunami victims are psychologically more distressed in the long term than non-tsunami victims, and to assess the extent of traumatization or any other mental disorders from which the victims of the tsunami still suffer, the author used two questionnaires: the "Impact of Event Scale – Revised (IES-R)" (Weiss & Marmar, 1996) and the "Brief Symptom Inventory (BSI)" (Derogatis, 1993). In both of these scales, the subjects were requested to check boxes from 0 to 4 to indicate their level of suffering from certain symptoms during the last seven days.

In order to find protective factors that may have helped the tsunami victims overcome their traumatization, she administered the "Sense of Coherence Scale (SOC scale)" (Antonovsky, 1987) to assess whether the people could find meaning in the event; "The Brief COPE Scale" (Carver, 1997) to determine which methods of coping with the tragedy were effective; and the "Social Support Questionnaire (Sozu K-22)", Sommer & Fydrich, 1989) to determine to what extent social support reduced the effects of the traumatic event.

7.4.1 Impact of Event Scale-Revised

The characteristic core of PTSD symptomology includes the distressing oscillation between intrusion and avoidance (Horowitz, 1978). Intrusive symptoms typically consist of nightmares and unwanted thoughts and memories of the event, whereas avoidance symptoms are characterized by a conscious effort to avoid talking or thinking of the event (see 2.3.2). Emotional numbing can also be part of the avoidance symptoms. According to Horowitz (1978), when an event falls outside normal expectations, as in a traumatic event, the information will be held in full iconic form in short-term memory for further conscious processing. The emotions that can be evoked by the appraisal of a traumatic event, which challenges basic assumptions (schemata), can be quite powerful. Thus, avoidance may be a normal way of coping with the threat of overwhelming emotion, and the individual will "dose" themselves with amounts of processing with which they are comfortable.

Following this conceptual framework, Horowitz et al. developed a simple but powerful self-report measure for assessing the extent of symptomatic response: the Impact of Event Scale (IES). The items on the IES were developed from statements most frequently used to describe episodes of distress by people who had experienced recent life events (Horowitz et al., 1979). The original IES was published before the appearance of the formal diagnostic criteria for PTSD in the DSM-III manual (American Psychiatric Association, 1980) and the subscales intrusion and avoidance described what was later called the B and C criteria of the PTSD diagnosis (Weiss, 2007). The hyperarousal symptoms presented in the D criterion were not included, such as hypervigilance, angry outbursts, and exaggerated startle response. When Horowitz et al. (1979) originally tested the reliability of the scale, they found two homogeneous clusters of items corresponding to intrusion and avoidance, with Cronbach's alpha of the intrusion subscale of 0.78, an alpha coefficient of 0.82 for the avoidance subscale, and an alpha coefficient for the total scale of 0.86. They also reported a test-retest reliability of $r = 0.89$ for intrusion and $r = 0.79$ for avoidance with a time interval of one week between tests. A correlation of .42 ($p<.01$) between subscales was reported, thus suggesting that the two variables are associated but do not measure identical constructs. In 1982, Zilberg, Weiss, and Horowitz investigated the internal consistency of a pooled patient and non-patient sample and found Cronbach's alpha values ranging from 0.86 and 0.90 across three points in time. Even though the IES does not correspond exactly to the DSM-IV criteria of PTSD, the scale is a very widely used self-report measure for traumatic stress symptoms. In May 2001, the PILOTS on-line database reported its use in 1,147 studies and a search of PsychINFO uncovered 515 citations of empirical studies (Weiss, 2007). The IES has been shown to differentiate between those who receive a diagnosis of PTSD and those who do not (Bryant & Harvey, 1996c), but it is not clear to what extent the IES is specific to PTSD. It was not developed to assess PTSD per se (Joseph, 2000).

In order to assess the full range of PTSD symptoms, including hyperarousal symptoms, Weiss and Marmar (1997) developed the Impact of Event Scale – Revised. A set of seven new items was developed, piloted, and then added to the original seven intrusion and eight avoidance items, to comprise a scale of 22 items. The revised version of the scale contains eight intrusion and eight avoidance items derived from the original IES, and adds six items assessing hyperarousal. The original structure of the scale was maintained as well as the original answering scheme of 0 = Not at all, 1 = Rarely, 3 = Sometimes, and 5 = Often. In both versions of the IES, the time frame of reference is the past seven days. In the IES-R, however, the respondents were asked to report the degree of distress of the symptoms in the past seven days instead of the frequency of symptoms. On the basis of experience, the response format of the IES-R was later modified to a 0–4 response format with equal intervals: 0 = Not at all, 1 = A little bit, 2 = Moderately, 3 = Quite a bit, and 4 = Extremely.

These modifications brought the IES-R in parallel format to the SCL-90-R (Derogatis, 1993), allowing for a direct comparison (Weiss, 2007). Finally, the authors recommend that the scoring system be modified: rather than computing a sum of the item scores, they suggest using the mean of non-missing items (Weiss & Marmar, 1997).

In a meta-analysis by Sundin, Mardi, and Horowitz (2003), data was evaluated from 66 studies that used the IES to examine the psychological impact of a major life event over a period of 20 years. Results from the hierarchical regression analyses suggest that only a small proportion of variation in IES scoring was attributed to sample characteristics such as age, gender, and cultural differences. In contrast, both type of traumatic event and time elapsed since the event were strong predictors of stress reactions levels. These findings indicate the value of the IES as a general, reliable, and valid self-report measure for assessing psychological stress reactions after negative life events (Sundin et al. 2003).

Another example of the cross-cultural validity of the IES is the examination of the Chinese version of the IES (IES-C) regarding its psychometric properties in newly diagnosed Taiwanese oral cancer patients. The results showed that (1) the IES-C has satisfactory content validity and feasibility; (2) overall internal consistency (Cronbach's Alpha) was 0.91 with values of 0.91 and 0.81 for intrusion and avoidance, respectively; (3) overall test-retest reliability (three-day interval) was 0.97; (4) two clearly identified factors explained 55.97 percent of the variance; and (5) satisfactory construct validity was supported by both factor analysis and theoretically supported correlation analysis (significant correlations between the total IES-C scores and its subscales as well as the HADS anxiety and depression subscales) (Chen, Lai, Liao, & Lin, 2005).

There is even a validated Tamil and Sinhalese version of the abbreviated eight-item version of the IES. The psychometric properties of the Tamil version are validated for adolescents in Tamil Nadu, South India, who had been exposed to diverse traumatic events (Russell, Subramanian, & Russell, 2004). Cronbach's alpha coefficients were high, suggesting that the Tamil version of the IES-8 has satisfactory internal consistency (intrusion: alpha = 0.87, avoidance: alpha = 0.77, and total score: alpha = 0.90). Pearson correlations between the two subscales were high (r = 0.70). The convergent validity between the Tamil IES and the Child Behaviour Checklist PTSD scale was marginally acceptable (intrusion r = 0.29, avoidance r = 0.22, total r = 0.29). Although modest, all correlations were significant at the 0.01 level. A two-factor structure explaining 68.4 percent of the variance was found, and all eight items loaded onto the expected factor, with the exception of item 5 ("I tried not to talk about it"), which loaded onto the intrusion factor.

The Sinhalese version of the IES-8 was validated for an adult sample of 30 survivors of the Asian 2004 tsunami in Sri Lanka (John & Russell, 2007). The majority belonged to the fishing community. The internal consistency was high for the entire IES (alpha = 0.78), but was moderate to high for the subscales (intrusion: alpha = 0.43 and avoidance: alpha = 0.82). The content validity was also satisfactory, as the two subscales were relatively independent of each other, each of them representing a different type of reaction (r = 0.46). The criterion validity between the Sinhalese version of the IES-8 and the DSM-IV-TR diagnosis of PTSD, calculated with Spearman's correlations, was moderate to high (intrusion rho = 0.56, avoidance rho = 0.76, and total rho = 0.83). The construct validity was determined with factor analysis. A two-factor structure explaining 60.8 percent of the variance was found, and all items loaded onto the expected factor, with the exception of item 6 ("Pictures about it popped into my mind"), which cross-loaded onto both the intrusion and avoidance factor.

Critical Voices

The possible paradoxical nature of avoidance in relation to PTSD has been the subject of discussion. Generally, the IES is used to provide a measure of distress. However, some researchers argue that the IES functions more as a measure of (incomplete) cognitive processing that mediates between stimulus exposure and symptoms (Creamer, Burgess & Pattison, 1992). In this way, there is a conceptual confusion between viewing avoidance as a symptom of PTSD or as a coping strategy that mediates between the traumatic event and subsequent emotional state (Joseph et al., 1995). According to Creamer et al. (1992), intrusion precedes avoidance in the development of PTSD. Avoidance is viewed as an anxiety-reducing coping strategy used in response to highly arousing traumatic intrusions. Thus, the two subscales of the IES could be conceptualized differently. Intrusion may be more directly symptomatic of the traumatic experience, whereas avoidance is a coping response to these symptoms (Joseph et al., 1995).

Moreover, Kagee (2005) argues that some of the items on the IES are not valence free. If these items are answered affirmatively, this does not mean that an individual is traumatized. For example, the item "I avoided letting myself get upset when I thought about it or was reminded of it" may as well indicate an adaptive response to a disturbing event. Similarly, the item "I tried to remove it from my memory" is not necessarily a maladaptive response.

It has also been criticized that the intrusion and avoidance subscales of the original IES were not designed to represent independent constructs that are be distinct and unique from general distress (Larsson, 2000). The scale might assess general negative affectivity rather than any trauma-specific phenomenon, as it was originally constructed to measure "the current degree of subjective impact experienced as a result of a specific event" (Horowitz et al., 1979, p.209). Moreover, as a two-factor structure, if the IES could not always be found, this points to the multidimensionality of the construct. Andrews, Shelvin, Troop, and Joseph (2004) specified and estimated a series of alternative confirmatory factor analytic models of the IES for an English sample. The best fitting model was a four-factor model comprising intrusion, avoidance, numbing, and sleep disturbances, all of which are subsumed by a second-order factor: general distress.

The primary disadvantage of the IES-R is that the items are not directly linked to the DSM diagnostic criteria of PTSD, making direct comparisons difficult. On the other hand, diagnostic criteria tend to change with each new edition, and there are advantages of focusing solely on measuring the core constructs of intrusion, avoidance, and arousal that characterize traumatic stress, rather than strictly adhering to somewhat arbitrary diagnostic criteria (Creamer, Bell, & Failla, 2003). This is why the IES has provided an unchanging standard measure of post-traumatic stress and can be described as the "gold standard self-report measure in trauma research" (Joseph, 2000, p. 108). Moreover, the IES is based on Horowitz's higher-order two-factor model of trauma response, and a higher-order two-factor model does in fact provide the best fit structure to trauma-related reactions (Joseph, 2000).

International Versions of the IES-R

In a variety of countries and cultures, versions of the IES-R have been produced in different languages. The scale has been translated into Chinese (Wu & Chan, 2003), Japanese (Asukai, Kato, Kawamura, Yamamoto, Kishimoto, et al., 2002), French (Brunet, St-Hilaire, Jehel, & Kind, 2003), German (Maercker & Schützwohl, 1998), Greek (Mystakidou, Tsilika, Parpa, Galanos, & Vlahos, 2007), and Spanish (Baguena, Villarroya, Belena, Amelia, Roldan, & Reig, 2001), and non-published versions exist in Dutch, Italian, Norwegian, and Persian (Weiss, 2007). The internal consistency of these translated scales is sufficient and in the range of 0.78 to 0.90. The data regarding test-retest stability were also consistent and encouraging (Weiss, 2007).

However, when testing the underlying factor structure of the scales, a three-factor solution was not always found. For example, the Chinese version data yielded a single factor that accounted for 45 percent of the variance (Wu & Chan, 2003). In the French version, both a two- and three-factor solution were found. The two-factor solution comprised an avoidance factor and intrusion-arousal factor (Brunet et al., 2003). This result was similar to the structure of the Spanish version by Baguena et al. (2001), which they labeled 'intrusion-hyperactivation and avoidance'. The Japanese IES-R scale was subject to a clustering method similar to factor analysis, and a three-factor solution was chosen beforehand. The three clusters are: intrusion-hyperarousal, avoidance, and numbing/sleep/cognitive distress (Asukai et al., 2002). The Greek version of the IES-R, on the other hand, showed the intended three-factor structure of intrusion, avoidance, and hyperarousal, with Cronbach's alpha values of 0.72, 0.77, and 0.85, respectively, explaining 57.26 percent of the variance (Mystakidou et al., 2007).

Psychometric Properties of the IES-R

Weiss and Marmar (1996) report psychometric data from two samples at two points in time: emergency personnel exposed to a freeway collapse, and workers from the 1994 Northridge earthquake. The IES-R showed high internal consistency, with coefficient alphas ranging from 0.87 to 0.92 for intrusion, 0.84 to 0.85 for avoidance, and 0.79 to 0.90 for hyperarousal. Test-retest correlation coefficients ranged from 0.57 to 0.94 for intrusion, 0.51 to 0.89 for avoidance, and 0.59 to 0.92 for hyperarousal. A principal components factor analysis with Varimax rotation revealed a strong single factor accounting for 49 percent of the variance. The authors suggest an explanation: not all respondents experienced high, or even medium, symptom levels. They conclude that the validation of the three clusters of PTSD symptoms remains to be adequately documented (Weiss & Marmar, 1996). In spite of the wide usage of the IES-R as a psychometric measurement and its translation into many different languages, surprisingly little has been published on the psychometric properties and construct validity of the scale (Creamer et al., 2003).

Creamer et al. (2003) investigated the psychometric properties of the IES-R in Australian male Vietnam veterans. The scale showed high internal consistency (alpha = 0.96). However, confirmatory factor analysis did not support a three-factor solution. Rather, exploratory factor analysis suggested that either a single (accounting for 56% of the variance), or two-factor solution (intrusion/hyperarousal and avoidance), provided the best account of the data. The correlation between the IES-R and the PTSD Checklist was high (r=0.84). Although neither the IES nor the IES-R were designed to make a categorical diagnosis, it is common for cut-off scores to be quoted in the literature. Thus, optimal cut-off scores for the IES-R were examined by using the recommended cut-off of 50 on the PTSD Checklist. Against this checklist, the highest overall diagnostic power (0.88) was achieved with a cut-off of 1.5 (equivalent to a score of 33) on the IES-R, providing a sensitivity of 0.91, a specificity of 0.82, positive predictive power of 0.90, and negative predictive power of 0.84 (Creamer et al., 2003). This is only slightly more conservative than the cut-off of 30 (an item mean of 1.4) proposed by Asukai et al. (2002).

In a study with motor vehicle accident victims in the United States, Beck, Grant, Read, Clapp, Coffey, Miller, et al. (2008) examined the factor structure, concurrent validity, and discriminative validity of the IES-R. Results supported the three-factor structure of the IES-R (intrusion, avoidance, and hyperarousal) with adequate Cronbach's alpha coefficients for each subscale. Significant correlations between each of the IES-R subscales and other PTSD, anxiety and depression measures (e.g. CAPS, STAI, BDI) were found. Cross-measure correlations for a specific subscale were also significantly higher than cross-measure correlations between two different subscales, providing support for the concurrent validity. The postulated discriminative validity was also supported, as significant differences were found between individuals with and without PTSD on each subscale. Finally, a discriminant function analysis (DFA) was conducted to examine the specificity and sensitivity of the IES-R subscales in identifying individuals with and without PTSD. The three subscales loaded on one function, which correctly classified 69.2 percent of the sample. Sensitivity was 74.5 and specificity was 63.1.

In a sample of refugees in Stockholm, Sweden, discriminant analysis was performed to find a cut-off score on the IES-R to discriminate between respondents suffering from clinically significant PTSD and respondents who did not suffer from it. ROC curves showed that a cut-off score of 65 was the most appropriate, with a sensitivity of 0.72 and a specificity of 0.71 (Söndergaard, Ekblad, & Theorell, 2003). In a Japanese study with psychiatric nurses being subject to violence by patients, the cut-off score of 24/25 was used (Inoue et al.).

7.4.2 Brief Symptom Inventory

The Brief Symptom Inventory (BSI) (Derogatis, 1993) is a 53-item self-report inventory assessing the psychological symptom status of psychiatric patients as well as individuals who are not patients. The participants are asked to what degree they have been suffering from these somatic and psychological symptoms for the past seven days; from 0 = "not at all" to 4 = "extremely". Test administration normally takes less than ten minutes, with one to two minutes devoted to the introduction of the instrument. Instructions preceding the administration of the inventory are brief and worded in simple language, as are the test items (Derogatis & Melisaratos, 1983). The BSI is an abbreviated form of the Symptom Checklist SCL-90 (Derogatis, 1993). Depending on the requirements of diverse psychological, medical, or pharmaceutical research studies, the SCL-90 itself underwent several changes until it culminated in the BSI as a more convenient short form. Many researchers and participants felt that the SCL-90, with its 90 items, took too long to administer. The BSI is a simple and cost-effective psychometric assessment method that measures a broad range of psychopathological states (Derogatis & Coons, 1993). When developing the BSI, Derogatis (1993) found five to six items to be sufficient for each of the nine subscales, so only the highest loading ones were chosen. The subscales measure nine different psychopathological syndromes. By the 1970s, Derogatis and Cleary (1977) had already validated the nine scales of the SCL-90 in a sample of 1002 psychiatric outpatients and 423 inpatients. When the hypothetical dimensions were matched against the empirical ones, eight out of the nine dimensions were judged to present a good theoretical agreement and some fared better than others. The only exception was the subscale "psychoticism," with only a moderate agreement. Thus, these findings confirm the construct validity of the instrument. The nine subscales are as follows:

1. Somatization (SOM)
2. Obsessive-compulsive (0-C)
3. Interpersonal sensitivity (I-S)
4. Depression (DEP)
5. Anxiety (ANX)
6. Hostility (HOS)
7. Phobic anxiety (PHOB)
8. Paranoid ideation (PAR)
9. Psychoticism (PSY)

Additionally, four items not belonging to a specific subscale assess loss of appetite, sleep disturbances, suicidal ideation, and feelings of guilt. These symptoms load onto several items, but are not unique to any of them. As they represent important vegetative and other clinical indicators, they were retained in the instrument. Three global indices were developed in order to be able to assess psychological distress in a more flexible way: Global Severity Index (GSI), Positive Symptom Distress Index (PSDI), and Positive Symptom Total (PST). The GSI combines information on the numbers of symptoms and the intensity of perceived distress. The PSDI is a pure intensity measure, in a sense 'corrected' for the number of symptoms. It functions very much as a measure of response style, indicating whether the patient is exaggerating or understating distress in his or her manner of reporting. The PST is simply a count of the symptoms that the patient reports (Derogatis & Melisaratos, 1983).

Reliability

The original author (Derogatis, 1993) examined the internal consistency (Cronbach's alpha) of the scales in a sample of psychiatric outpatients, and the alpha coefficients ranged from 0.71 (psychoticism) to 0.85 (depression). Boulet and Boss (1991) evaluated the BSI in a sample of 501 forensic-psychiatric patients, and the alpha coefficients ranged from 0.75 (psychoticism) to 0.98 (depression). Derogatis (1993) found a test-retest reliability of alpha coefficients ranging from 0.68 (somatization) to 0.91 (phobic anxiety) when retesting a sample of 60 psychiatric outpatients over a two-week time interval.

Validity

In order to assess the convergent validity of the BSI, Derogatis and Melisaratos (1983) correlated the subscales of the BSI with the corresponding measures of the Minnesota Multiphasic Personality Inventory (MMPI) and found sizable correlations with like constructs. Although the overall magnitude of the correlations was reduced on several dimensions, the authors concluded that the reduction in length of the SCL-90-R symptom scales did not compromise their validity. However, Boulet and Boss (1991) criticized these findings, stating that the nine BSI dimensions were found to correlate both with analogous and non-analogous measures of the MMPI, thus revealing a limited convergent validity and poor discriminant validity for the instrument.

Convergent validity was also examined by correlating the BSI scales with other psychometric constructs such as social support. As expected, those participants with low perceived social support (measured with the Social Support Questionnaire by Fydrich, Sommer, Menzel, & Höll, 1987) had the highest GSI scores on the BSI and the highest score on the subscale "depression" ($r = 0.63$). Face validity is given, as the BSI is the offspring of the SCL-90, a scale constructed through a long process of interplay between expert opinions and patient interviews (Franke, 2000).

In order to assess the construct validity of the BSI, Boulet and Boss (1991) performed a principle components analysis with their sample of 501 forensic-psychiatric patients. Only one main factor was found. The first factor accounted for 71 percent of the variance, and a second factor for 5.9 percent of the variance. Regarding this sample, not much additional information is gained by splitting the instrument into nine separate psychopathology subscales. Moreover, these authors found that most items do not correlate highest with their subscale constructs. Only 29 out of 49 items load highest on their corresponding subscale. The total score of the BSI was highly correlated with each dimension score, suggesting that a single global score might be used as an index of psychopathology, at least for this sample, which was extreme in terms of psychopathology.

In several studies, the BSI was used as a diagnostic measurement to assess psychological distress after a traumatic event. For example, Collings (1997) found that female university students in South Africa who were victims of sexual assault had higher scores on all nine subscales compared to a non-victim group. A study with emergency workers both professionally and personally affected by Hurricane Andrew revealed a mean of 0.58 on the BSI Global Severity Index, which indicates more distress than normal (Jenkins, 1997). Moreover, Ouimette, Ahrens, Moos, & Finney (1998) found that substance abuse patients with PTSD had higher scores on the BSI than substance abuse patients without PTSD.

International Versions of the BSI

The BSI has been used in different countries and for different ethnic groups. Thus, it has been translated into several different languages such as German (Franke, 2000), Spanish (Acosta, Nguyen, & Yamamoto, 1994), Polish (Arioan & Patsdaughter, 1989), and Filipino (Aroian, Patsdaughter, Levin, & Gianan, 1995). Intercultural differences were investigated by several authors. For example, Aroian et al. (1995) investigated the use of the BSI with Polish and Filipino immigrants and its parent version, the SCL-90, with Irish immigrants. In all three samples, sufficient internal consistency was found, which shows that the BSI is a relatively reliable and cross-culturally valid measure of psychological distress. The only subscale not showing high alpha ranges in any sample was "psychoticism." Iwamasa and Kooreman (1995), as well as Lippincott and Mierzwa (1995) investigated the differences between Asian-American and Euro-American college students and found that the level of acculturation influenced the reporting of distress on the BSI. The students with higher levels of acculturation showed symptoms similar to their Euro-American counterparts. Moreover, students with higher levels of acculturation showed higher levels of somatization. The authors conclude that these students, especially female ones, experience more distress, as they try to fit into both cultures and express their distress somatically.

7.4.3 Sense of Coherence Questionnaire

The SOC questionnaire (Antonovsky, 1987) was chosen for this study because it was designed to be culture free. Apparently similar constructs such as self-efficacy, hardiness, and locus of control are essentially culture bound. Beneath the latter constructs lies the assumption that the environment in which people live poses problems to which the culturally appropriate responses are active mastery, internal, and control responses. However, in different cultures or subcultures, or in different situations, successful coping responses may differ. According to Antonovsky (1992), a high SOC score can legitimately be obtained in any cultural setting. To believe that stressors can be comprehended and managed (with others' help, if needed) and are worthy of engagement is always culturally acceptable.

By 1993, the SOC scale had already been used in 14 languages: Afrikaans, Czech, Dutch (Flemish), English, Finnish, German, Hebrew, Norwegian, Rumanian, Russian, Serbian, Spanish, Swedish, and Tswana. All social classes had also been represented (Antonovsky, 1993). In a systematic literature review, Eriksson and Lindström (2005) found that between 1992 and 2003, the SOC was used in at least 32 languages and in 32 countries with at least 15 different versions of the questionnaire. The SOC questionnaire has been used in both Western countries and countries such as Thailand, China, Japan, and South Africa. Therefore, the SOC questionnaire seems to be a cross-culturally applicable instrument.

The means of the SOC-29 range from 100.5 (SD 29.50) to 164.50 (SD 17.10) points (124 studies) and the SOC-13 from 35.39 (SD 0.10) to 77.60 (SD 13.80) points (127 studies) (Eriksson & Lindström, 2005).

Reliability

In 26 studies in different languages and cultures using SOC-29, the Cronbach's alpha measures of internal consistency ranged from 0.82 to 0.95. The alphas of 16 studies using the SOC-13 ranged from 0.74 to 0.91. The relatively few test-retest correlations show considerable stability, for example, 0.54 over a two-year period among Israeli retirees (Antonovsky, 1993). According to Eriksson and Lindström (2005), in 124 studies using the SOC-29 the Cronbach's alpha ranges from 0.70 to 0.95. The alpha values in 127 studies using the SOC-13 range from 0.70 to 0.92, and in 60 studies using a modified SOC scale, they range from 0.35 to 0.91. Test-restest correlation also shows stability and ranges from 0.69 to 0.78 (1 year), 0.64 (3 years), 0.42 to 0.45 (4 years), 0.59 to 0.67 (5 years) to 0.54 (10 years). After ten years, the SOC seems to be comparatively stable, but not as stable as Antonovsky (1987) initially assumed. SOC seems to increase with age (Eriksson & Lindström, 2005).

Validity

The content validity was obtained by letting three colleagues independently agree on an item referring to only one of the three SOC components before including the item in the scale. Moreover, a post hoc examination of the scale suggested face validity in that it was adequately representative of the theoretical construct. The lack of a relationship between SOC and an interpersonal support measure suggests the discriminant validity of the SOC scale; this means that the SOC functions independently of socially-based generalized resistance resources (Antonovsky, 1993).

One methodological danger could be the contamination of the independent and dependent variable, that is, the variables are actually aspects of the same phenomena. The SOC questionnaire could measure the same variables as health questionnaires. This danger can be avoided by focusing on the pathogenic variable as the dependent variable, but this does not give insight into health variables. That is why Antonovsky scrutinized every item of the SOC questionnaire to avoid possible contamination with the concept of health (Antonovsky, 1992).

Criterion validity is clearly visible, as the SOC correlates significantly with constructs to which it is supposed to be correlated according to theory. Some examples are: internal locus of control, self-esteem, hardiness, trait anxiety, stress perception, global health index, and quality of life. SOC is positively related to problem-focused coping, but negatively related to emotion-focused coping. It is worth mentioning that a large sample of minority, homeless, and drug-addicted women had the lowest SOC-13 score, whereas kibbutz members and American university faculty had the highest scores, a phenomenon that also supports the validity of the scale (Antonovsky, 1993).

Factor Structure of the SOC scale

Even though the construct of SOC is comprised of the three components: comprehensibility, manageability, and meaningfulness, according to Antonovsky (1993), there is no basis for deriving distinguishable subscores for these components. A factor analysis on the original Israeli national sample demonstrated that no separate meaningful factors could be identified. Up to that date, no published studies reported a factor analysis, although several unpublished studies showed that a principal component analysis produced only one global factor (eigenvalue 12.45, 42.9% of the variance). It can be concluded that the content of the items can be categorized into separate factors, but the manageability, meaningfulness, and comprehensibility dimensions are present in each factor. In this way, the SOC scale can be viewed as an instrument measuring a global construct in which certain aspects can be highlighted.

However, according to Eriksson and Lindström (2005), the factorial structure of the scale is not completely clear. Some studies confirmed the one-factor solution proposed by Antonovsky (1987) (Frenz, Carey, & Jorgensen, 1993). However, in other studies, the factorial structure of the SOC seems to be multidimensional rather than unidimensional. For example, among 300 Swedish students, a five-factor solution was reported: meaning in life, control of feelings, negative feelings, trust/distrust, and changing future, explaining 53 percent of the variance (von Bothmer & Fridlund 2003). A three-factor solution that differs from Antonovsky's, was proposed by Sandell, Blomberg, and Lazar (1998) on Swedish patients and controls.

Regarding the criterion validity, Eriksson and Lindström (2005) found the correlation of SOC with measures of health to be slight to good (slight meaning 0.20 to 0.35 and good meaning $r = 0.65$ to 0.85). The studies reviewed used instruments such as the general health questionnaire, health index, Hopkin's Symptom Checklist, or mental health inventory explaining at highest 66 percent of the variance in SOC. The most striking finding was the relatively high negative correlation with anxiety and depression, and the strong positive correlation with optimism and self-esteem.

7.4.4 Brief COPE Scale

The way in which individuals cope with traumatic events and which resources they can mobilize should also be part of the diagnostic process. The focus should not be solely on dysfunctional behavior, but also on how the individual adapts to the traumatic experience (Butollo et al., 1998). One questionnaire measuring coping strategies is the COPE created by Carver et al. (1989). In COPE, coping strategies such as planning behavior, seeking assistance, acceptance, denial, distraction, and substance use are assessed. Even though the distinction between problem-focused and emotion-focused coping is widely acknowledged, many authors, including Carver et al. (1989), felt that this was too simple. They believe that both coping functions have to be subdivided because there are a variety of distinct ways to solve problems or regulate emotions. Thus, after several generations of its development, the final version of the COPE scale contains 13 scales with four items each (the authors now distribute a version with two additional experimental scales). When developing the scale, the authors used rational construction procedures to compose diverse scales, which were then tested in a sample of 978 undergraduates. The students were asked what they usually did when they were under considerable stress. A factor analysis produced almost the same structure as hypothesized (Carver et al., 1989). However, it is not reported how many items were initially used, and where the 13 scales come from after only 11 meaningful factors were extracted (Schwarzer & Schwarzer, 1996).

The 13 scales are as follows:

1. Positive reinterpretation and growth
2. Mental disengagement
3. Focus on and venting of emotions
4. Seeking social support for emotional reasons
5. Active coping
6. Denial
7. Turning to religion
8. Behavioral disengagement
9. Restraint coping
10. Seeking social support for instrumental reasons
11. Acceptance
12. Suppression of competing activities
13. Planning

The above scales were found to have satisfactory psychometric properties, with only one scale having an alpha coefficient below 0.60. The exception was the mental disengagement scale with somewhat ambiguous item phrasing (Carver et al., 1989), although evidence for validity is provided (Carver, Scheier & Prozo, 1992).

According to Carver et al. (1989), there is no such thing as an "overall" coping score on this measure. Therefore, they recommend no particular way of generating a dominant coping style for a given person. Rather, they recommend looking at each scale separately to determine its relation to other variables. Another alternative is to create second-order factors from among the scales and use the factors as predictors. Carver et al. (1989) also conducted a second-order factor analysis (omitting the alcohol and substance abuse items) and found four second-order factors: one factor was composed of active coping, planning, and the suppression of competing activities. The second one was composed of seeking social support (both scales) and focused on emotion. A third factor was composed of denial and both mental and behavioral disengagement. The fourth factor consisted of acceptance, restraint coping, and positive reinterpretation and growth. However, according to Carver et al. (1989), each researcher should use his or her own data to determine the composition of the higher order factors, as different samples exhibit different patterns of relations. Moreover, depending on the type of research question, the COPE scale can either measure dispositional or situational coping. The formats differ in their verb forms: the dispositional "trait version" is in the present tense (I usually do this…) and the "state form" is in the present perfect (I have been doing this...), referring to a concrete event.

Another variable that previous research has linked to variations in coping is the controllability of the event. When situations are controllable, active strategies are predominately used, and when situations are uncontrollable, alternative strategies predominate (Folkman & Lazarus, 1980). In Carver et al. (1989) study of college undergraduates, it was likewise found that active coping, planning, suppression of competing activities, and seeking social support for instrumental reasons were positively associated with optimism and the feeling of being able to do something about the situation. Those students who reported viewing the situation as something that they had to be used to reported higher levels of both acceptance and denial.

A 30-item Hebrew version based on the original scale developed by Carver et al. (1989) was used to test the influence of coping behavior on test performance in a sample of Israeli undergraduate students (Cohen et al., 2008). The items were divided into 15 coping strategies with two items per strategy. A factor analysis with Varimax rotation showed a structure of four factors, similar but not identical to the one reported by Carver et al. (1989). This analysis was then used to create a problem-focused scale consisting of active coping, planning, and suppression (alpha = 0.67); an emotion-focused scale, including seeking emotional support, instrumental support, and ventilation (alpha = 0.69); and an avoidance scale, including mental disengagement, behavioral disengagement, and denial (alpha = 0.70). The fourth factor only included the acceptance strategy, which was not included in the analysis. Humor, religion, and alcohol were also not included in the analysis. The results of the study showed that problem-focused coping contributed positively to performance on the test, and avoidance coping adversely affected test grades, mediated by test anxiety.

The Brief COPE

In 1997, Carver constructed an abbreviated version of the COPE scale, because of the length and redundancy of the full instrument and of the overall time burden of the assessment protocol. In choosing which items to retain for this version (which has only two items per scale), the author was guided by strong loadings from previous factor analyses, and by item clarity and meaningfulness to the patients in a previous study. Some of the original scales were given slightly different names, and one scale that was not part of the original inventory (a two-item measure of self-blame) was added because that response has been important in some earlier work. The Brief COPE thus consists of 28 items that assess the frequency with which respondents engage in certain coping behaviors. Item scores range between 1 = "not at all" to 4 = "a lot". The 14 subscales are as follows:

1. Self-distraction (items 1 and 19)
2. Active coping (items 2 and 7)
3. Denial (items 3 and 8)
4. Substance use (items 4 and 11)
5. Use of emotional support (items 5 and 15)
6. Use of instrumental support (items 10 and 23)
7. Behavioral disengagement (items 6 and 16)
8. Venting (items 9 and 21)
9. Positive reframing (items 12 and 17)
10. Planning (items 14 and 25)
11. Humor (items 18 and 28)
12. Acceptance (items 20 and 24)
13. Religion (items 22 and 27)
14. Self-blame (items 13 and 26)

In a study with Hurricane Katrina evacuees, Sprang and LaJoic (2009) used the Brief COPE (Carver, 1997) to assess the coping strategies of the respondents. In their study, higher order factors were identified using principle components analysis with Varimax rotation to reduce the subscales to a more useful number. As a result of this analysis, seven factors were found accounting for 72 percent of the variance. Factor one consisting of 6 items explained 30 percent of the variance and was titled "avoidant coping," with Cronbach's alpha being 0.88. Avoidant coping is comprised of denial, behavioral or emotional disengagement, or activities opposed to active, task-oriented coping (e.g. self-blame). The other factors were not used in this study and further explanation was not given, as the authors investigated only the mediating role of avoidant coping between exposure and the onset of PTSD symptoms.

7.4.5 Social Support Questionnaire (Fragebogen zur Sozialen Unterstützung, F-SOZU; SOZU-K-22)

The original F-SOZU (Fydrich et al., 1987) was constructed in the German language to assess the individual's perceived social support. This perception of social support is assessed using three main scales: emotional support, tangible support, and social integration. An additional scale measures perceived social strain. The respondents were asked to answer the items on a five-point Likert scale, ranging from 0 = "I strongly disagree." to 5 = "I strongly agree."

In 1989, Sommer and Fydrich developed a longer version of the F-SOZU with 54 items to solicit additional information. On an additional scale, respondents list individuals in the social environment who are perceived either as a source of support or a source of strain. This longer version of the Social Support Questionnaire consists of two parts: Part A assesses the individual experience of social support with four scales (emotional support, tangible support, social integration, and social strain). Emotional support (16 items) measures, for example, being accepted and liked by others, being able to share positive and negative feelings, or experiencing empathy. Tangible support (9 items) measures practical support with every day problems, such as borrowing an object or receiving advice. Social integration (13 items) measures the availability of a social network, for example, people with similar interests, or people with whom one can engage in joint activities. Social strain (12 items) measures the extent of feeling criticized, rejected, or restricted by others. In the 22-item version of the F-SOZU, the first three scales of part A assessing subjectively perceived social support are summarized into an overall "perceived social support" score. However, in the longer version, these scales can be evaluated separately. In Part B of the F-SOZU, quantitative aspects of the social network are assessed, such as the number of people in the social environment who are perceived a source of support or as a source of strain (10 items). Additionally, sociodemographic data regarding social interactions are assessed, such as marital status, children, number of people living in the household, work situation, membership in a club or other social group, or frequency of social contacts in everyday life.

Psychometric Properties

In a study by Sommer and Fydrich (1991) with clinical and non-clinical subjects, the reliability and validity of the questionnaire was tested and proven to be satisfactory. The internal consistency of the F –SOZU scales ranged from Cronbach's alpha coefficients of 0.73 to 0.94 in different samples (samples with university students, the general population, relatives of schizophrenic patients, women with breast cancer, psychosomatic patients, and schizophrenia patients). The overall perceived social support score for the SOZU-K-22 ranged from Cronbach's alpha = 0.79 to 0.92, tested in 12 samples with a total of 909 respondents. Retest reliability was tested in a sample with female breast cancer patients and alpha values ranged from 0.52 to 0.65 (Fydrich et al., 1987).

The content validity was given, as the items of the F-SOZU were carefully derived from the construct "social support." Additionally, nineteen experts in the field of psychological research estimated the content validity of each item, and only those items chosen rated as having high content validity were chosen.

The three main scales of the questionnaire (emotional support, tangible support, and social integration) showed significant positive correlations with each other (in the pooled sample, r ranged from 0.61 to 0.72). This is not surprising, as a person who feels socially integrated will also score high on emotional and tangible support. Nevertheless, the correlations are low enough for the three scales to still distinguish between different constructs. The social strain scale showed significant negative correlations with the three other scales, especially the social integration scale.

The convergent validity also proved to be high, as social support showed positive correlations with social competence and life satisfaction, as well as negative correlations with feelings of powerlessness, physical complaints, social anxiety and psychopathology (Fydrich et al., 1987). For the "social strain" scale, the correlations were reversed; emotional support and social integration showed higher correlations than tangible support.

In a principle component factor analysis with Varimax rotation, the postulated four-factor structure was confirmed. The first factor was found to be compatible with the "emotional support" scale, as all items of this scale, with the exception of one, loaded onto this factor. The second factor consisted of eight of the nine items for tangible support, three items of emotional support, and one item for social integration, so it can be interpreted as "tangible support." The third factor comprised 11 out of 12 items for social strain, as well as one item each for emotional support and tangible support. In this way, it can be considered equal to "social strain." The fourth factor is less likely to be compatible with "social integration," as only four of the social integration items loaded onto this factor. However, the other nine items of the social integration scale did not strongly load onto any other factor. All four factors explained 37.5 percent of the variance (Fydrich et al., 1987).

8 Results

In order to determine whether tsunami victims are still suffering from traumatic stress symptoms two and a half years after the tsunami, t-tests comparing the mean test scores of the two groups were used. Moreover, to assess the role of protective factors such as SOC, social support, and coping strategies, correlation analyses and regression analyses were used. The influence of demographic variables on the extent of traumatic stress was investigated using cross-tabulations. Finally, a path-analysis model explored multiple pathways and mediating factors.

8.1 Exploratory Factor Analyses

In order to find reliable subscales for the five questionnaires used in this study, factor analysis was used for each of the instruments. As the existing subscales were already validated by the respective authors in samples drawn in Western industrialized cultures, it was necessary to find new subscales in accordance with the data of a sample of villagers in Kerala, South India. Through structure-analyzing procedures, the interrelationships between the items of the instruments could be identified, and then through data reduction be grouped into dimensions or factors. A factor is a linear combination or cluster of related observed variables that represents a specific underlying dimension of a construct. The factors should be as distinct from each other as possible. The goal of factor analysis is to arrive at a reduced set of factors that summarizes and describes the structural interrelationships between the variables.

8.1.1 A Justification for the Choice of Factor Extraction and Rotation Methods

Factor Extraction Method

In all of the factor analyses performed for this thesis, the factor extraction method used was principle axis factoring (PAF). In another commonly used factor extraction method, the principal component analysis (PCA), the unique variance of a particular item, including its errors of measurement, is not factored out. It is assumed that the items included in the analysis can be perfectly calculated by the extracted components. PAF, however, separates the variance that the items share from this unique variance. According to PAF, variance in a given item is explained by a small number of underlying factors as well as variance that is unique to the item. Thus, the factors extracted in PAF are not mere linear combinations of the items being examined, as in PCA, but rather hypothetical factors that are estimated from the items being examined. In this way, the variance values of the factors are unique and intuitively make sense.

Rotation Method

In the factor analyses performed, Varimax (Kaiser, 1958, 1959) orthogonal rotation was used. An assumption underlying an orthogonal rotation is that the subscales formed from the factors are independent of each other. In an orthogonal rotation, the factors are maintained at right angles to one another and as a result, these newly rotated factors are also uncorrelated. Varimax is the most commonly used orthogonal rotation. A Varimax solution is easily interpreted and provides relatively clear information about which items correlate most strongly with a given factor. Factor scores generated for each individual are also more interpretable because the explained variances among the factors do not overlap and are therefore independent of each other.

Selecting the Number of Factors to Retain

Since the goal of factor analysis is to reduce the number of factors or components in order to maximize the amount of variance explained with the fewest number of factors or components, the following guidelines are used to determine the number of factors to extract: One criterion is that the eigenvalues of the factors must be larger than 1.00. Another criterion is that the cumulative percentage of variance extracted by successive factors has reached a certain percentage, e.g. 75 to 80 percent. Each factor should explain at least 5 percent of the variance. Finally, the number of factors can be determined by examining the scree plot. In the scree plot, there is a distinct break between the steep slope of the larger eigenvalues and the trailing off of the smaller ones. To find the exact location of the break, a straight line is drawn with a ruler through the lower values of the plotted eigenvalues. That point where the factors curve above the straight line identifies the number of factors (Cattell & Jaspers, 1967).

Internal Consistency of the Factors (Cronbach's Coefficient Alpha)

An important and widely used measure for assessing the internal consistency of a set of items is Cronbach's coefficient alpha (α) (Cronbach, 1951, 1984). It is a measure of reliability that represents the proportion of total variance in a given scale that can be attributed to a common source. It can also be defined as the estimated correlation of the given scale with another scale of the same length from the universe of possible items. Coefficient alpha has an advantage over split-half reliability in that it is not a single estimate of a scale's reliability, but rather the average of all possible split-half reliability coefficients.

8.1.2 Approach to Calculation of Factor-Based Scores

Once the number of factors have been determined, it is useful to generate a score for a particular respondent on each identified factor. These scores can be used in subsequent analyses. For example, once several subscales of a psychometric instrument have been identified, the researcher can use the generated composite scores to represent values for the factors and examine their relationship to demographic and psychosocial characteristics of the sample that have been measured or to examine the differences between selected subgroups of respondents.

8.1.3 Factor Analysis of the Impact of Event Scale – Revised (IES-R)

In the IES-R, a three-factor solution was postulated by Weiss and Marmar (1996), consisting of 'intrusion,' 'avoidance' and 'arousal' factors. However, other authors in other countries have found different factor solutions (see 7.4.1). This is why the author wanted to determine which factor solution would be adequate for this sample.

Before the performance of the PAF factor analysis for the IES-R, the author evaluated the correlation matrix according to Bartlett's test of sphericity and the Kaiser-Meyer-Olkin test. The size of the Kaiser-Meyer-Olkin (KMO) measure, which is a measure for the sampling adequacy, was 0.935. This is considered "marvelous," according to Kaiser (1974). Above 0.90 is "marvelous," in the 0.80s is "meritorious," in the 0.70s is just "middling," and less than 0.60 is "mediocre," "miserable" or "unacceptable." The KMO indicates the strength of the relationship among items. Bartlett's test of sphericity tests the null hypothesis that the correlation matrix is an identity matrix (i.e., that there is no relationship among items). Bartlett's test is a chi-square test, and larger values indicate a greater likelihood that the correlation matrix is not an identity matrix and that the null hypothesis will be rejected. Here the χ^2 value is significant, which means that the null hypothesis can be rejected and is some correlation exists between items.

Criteria for Extracting the Factors for the IES-R: the Eigenvalues, Percent of Variance Extracted, and the Scree Plot

According to the number of eigenvalues larger than 1.00, the 5 percent - criterion and examination of the scree plot, three factors could have been extracted (see Figure 1). However, the two-factor solution is the one chosen because it makes more sense intuitively (see Figure 2). The items are easier to interpret in a two-factor solution. Moreover, there should be at least two items in each factor. Therefore, the chosen two factors explain 56.3 percent of the variance. The factors are 'intrusion/arousal' with Cronbach's alpha being 0.937 and 'avoidance' with Cronbach's alpha being 0.873. The first factor explains 49.1 percent of the variance and the second factor 7.2 percent. A third factor would have explained only 5.4 percent more variance.

Thus, the two-factor solution corresponds to the factor solution of the original author of the IES (Horowitz, 1979). A two-factor solution of the IES-R was also found by other authors, for example, by Baguena et al. (2001) in the Spanish version. They labeled their factors 'intrusion-hyperactivation' and 'avoidance.' In the French version, both a two- and a three-factor solution were found. The two-factor solution comprised of an avoidance factor and a combined intrusion-arousal factor (Brunet et al., 2003). A single-factor solution was found in the Chinese version. Here, the data yielded a single factor that accounted for 45 percent of the variance (Wu & Chang, 2003). As the first factor of this dissertation study explains 49.1 percent of the variance and the other factor only 7.2 percent, one could argue that a single-factor solution might be sufficient. However, it would not be possible to differentiate between avoidance and intrusion symptoms and explore the association with other variables.

Descriptions and Interpretations of the Factors of the IES-R

The first factor or subscale of the IES-R is called "intrusion/arousal" and consists of 14 items (items 1, 2, 3, 4, 6, 9, 10, 14, 15, 16, 18, 19, 20, 21). These items were selected for the subscale because they had the highest correlations with the factor and they determine the highest score for Cronbach's coefficient alpha ($\alpha = 0.94$). These items are as follows:

1. Any reminder brought back feelings about it.
2. I had trouble staying asleep.
3. Other things kept making me think about it
4. I felt irritable and angry.
6. I thought about it when I didn't mean to.
9. Pictures about it popped into my mind.

10. I was jumpy and easily startled.
14. I found myself acting or feeling as though I was back at that time.
15. I had trouble falling asleep.
16. I had waves of strong feelings about it.
18. I had trouble concentrating.
19. Reminders of it caused me to have physical reactions, such as sweating, trouble breathing, and a pounding heart.
20. I had dreams about it.
21. I felt watchful or on-guard.

The second subscale "avoidance" consists of the eight items 5, 7, 8, 11, 12, 13, 17, and 22 and has Cronbach's α value of 0.87. The items are as follows:

5. I avoided letting myself get upset when I thought about it or was reminded of it.
7. I felt as if it hadn't happened or wasn't real
8. I stayed away from reminders about it.
11. I tried not to think about it.
12. I was aware that I still had a lot of feelings about it, but I didn't deal with them.
13. My feelings about it were kind of numb.
17. I tried to remove it from my memory.
22. I tried not to talk about it.

8.1.4 Factor Analysis of the Brief Symptom Inventory (BSI)

In the BSI (Derogatis, 1993), nine subscales or underlying factors were postulated. However, in several studies, this nine-factor solution was not confirmed (see 7.4.2). In a study with forensic-psychiatric patients by Boulet and Boss (1991), only one main factor was found. The first factor accounted for 71 percent of the variance and a second factor for 5.9 percent of the variance. The total score of the BSI was highly correlated with each dimension score, suggesting that a single global score might be used as an index of psychopathology, at least for this sample, which was extreme in terms of psychopathology. In order to find the appropriate factor solution for this sample of Indian fisherfolk, the author performed a PAF factor analysis for the BSI.

Bartlett's test of sphericity showed that sampling adequacy was 0.96 and thus 'marvelous,' and the Kaiser-Meyer-Olkin test revealed that the correlation matrix was not an identity matrix (see Table 5). Thus it was possible to proceed with the factor analysis.

Criteria for Extracting the Factors for the BSI: the Eigenvalues, Percent of Variance Extracted, and the Scree Plot

Eight factors with eigenvalues larger than 1.00 were found (which together explain 69.4 percent of the variance), and two of the factors explained more than 5 percent of the variance. According to the scree plot, three factors should be extracted (see Figure 3). The final solution decided on was three factors, so that each subscale consisted of at least two items (see Figure 4). The first factor explained 49.4 percent of the variance, the second factor 5.9 percent, and the third factor 3.1 percent. Altogether, the three factors explained 58.3 percent of the variance. Anxiety, depression, as well as somatization symptoms all load together on one factor, which is called 'anxiety/depression/somatization' (Cronbach's alpha = 0.94). The second factor is 'hostility' (Cronbach's alpha = 0.92) and the third one is 'interpersonal sensitivity' (Cronbach's alpha = .92).

Descriptions and Interpretations of the Factors of the BSI

The first subscale "anxiety/depression/somatization" consists of 31 items 1, 2, 5, 7, 8, 9, 11, 12, 15, 17, 18, 19, 23, 25, 26, 27, 28, 29, 30, 31, 32, 33, 35, 36, 37, 39, 42, 43, 45, 47, and 50 (Cronbach's $\alpha = 0.97$). The items are as follows:

1. Nervousness or shakiness inside.
2. Faintness or dizziness.
5. Trouble remembering things.
7. Pains in heart or chest.
8. Feeling afraid in open spaces.
9. Thoughts of ending your life.
11. Loss of appetite.
12. Suddenly scared for no reason.
15. Feeling blocked getting things done.
17. Feeling blue.
18. Feeling no interest in things.
19. Feeling fearful.
23. Nausea or upset stomach.
25. Trouble falling asleep.
26. Having to check and double-check what you do.
27. Difficulty making decisions.
28. Feeling afraid to travel on buses or train.
29. Trouble getting your breath.

30. Hot or cold spells.
31. Having to avoid certain things, places or activities because they frighten you.
32. Your mind going blank.
33. Numbness or tingling in parts of your body.
35. Feeling hopeless about the future.
36. Trouble concentrating.
37. Feeling weak in parts of your body.
39. Thoughts about death or dying.
42. Feeling very self-conscious with others.
43. Feeling uneasy in crowds.
45. Spells of terror or panic.
47. Feeling nervous when you are alone.
50. Feelings of worthlessness.

The second subscale "hostility" consists of 12 items: 24, 34, 38, 40, 41, 44, 46, 48, 49, 51, 52, and 53 (Cronbach's $\alpha = 0.921$). The items are as follows:

24. Feeling that you are watched or talked about by others.
34. The idea that you should be punished for your sins.
38. Feeling tense or keyed up.
40. Having urges to beat, injure or harm someone.
41. Having urges to break or smash things.
44. Never feeling close to another person.
46. Getting into frequent arguments.
48. Others not giving you proper credit for your achievements.
49. Feeling so restless you couldn't sit still.
51. Feeling that people would take advantage of you if you let them.
52. Feelings of guilt.
53. The idea that something is wrong with your mind.

The third subscale "interpersonal sensitivity"consists of ten items: (3, 4, 6, 10, 13, 14, 16, 20, 21, and 22) (Cronbach's $\alpha =$ o.924). The items are as follows:

3. The idea that someone else can control your thoughts.
4. Feeling others are to blame for most of your troubles.
6. Feeling easily annoyed or irritated.
10. Feeling that most people cannot be trusted.
13. Temper outbursts that you could not control.
14. Feeling lonely even when you are with people.
16. Feeling lonely.

20. Your feelings being easily hurt.
21. Feeling that people are unfriendly or dislike you.
22. Feeling inferior to others.

8.1.5 Sense of Coherence Scale (SOC Scale)

According to Antonovsky (1987), the SOC scale is comprised of the three factors comprehensibility, manageability, and meaningfulness, corresponding to the three components of the SOC construct. However, Antonovsky (1993) advises using only one general factor for meaningful statistical comparison because there is no basis for deriving distinguishable subscores for these components. A factor analysis on the original Israeli national sample demonstrated that no separate meaningful factors could be identified. Up to that date, no published studies reported a factor analysis, although several unpublished studies showed that a principal component analysis produced only one global factor (eigenvalues 12.45, 42.9 % of the variance). It can be concluded that the content of the items can be categorized into separate factors, but the manageability, meaningfulness, and comprehensibility dimensions are present in each factor. In this way, the SOC scale can be viewed as an instrument measuring a global construct in which certain aspects can be highlighted.

After performing a PAF factor analysis on this sample of Keralite fisherfolk, a three-factor solution was found. However, the items do not correspond to the original three factors of comprehensibility, manageability, and meaningfulness. Therefore, only the total score of the SOC Scale was considered as meaningful. Moreover, only one subscale showed sufficient internal consistency to be considered a valid subscale. The internal consistency of the subscales would have been alpha = 0.74, alpha = 0.60, and alpha = 0.23, respectively.

In addition, Bartlett's test of sphericity revealed a value of 0.72, which is 'middling,' according to Kaiser (1974). The Kaiser-Meyer-Olkin test led to a significant chi-square, meaning that the correlation matrix is not an identity matrix (see Table 6). Thus, the factor analysis could be performed.

Criteria for Extracting the Factors for the SOC Scale: the Eigenvalues, Percent of Variance Extracted, and the Scree Plot

Four factors have eigenvalues larger than 1.00, and all four factors explain more than 5 percent of the variance. The examination of the scree plot also suggests a four-factor solution (see Figure 5). However, only three factors have been extracted (see Figure 6) because only one item loaded on the fourth factor. All three factors explained 47.6 percent of the variance. The first factor explained 21.7 percent of the variance, the second factor 15.3 percent, and the third factor 10.6 percent.

Descriptions and Interpretations of the Factors of the SOC Scale

The first factor consists of five items: 6, 9, 11, 12, and 13 (Cronbach's α = 0.74). The items are as follows:

6. Do you have the feeling that you are in an unfamiliar situation and don't know what to do?
9. Does it happen that you have feelings inside you would rather not feel?
11. When something happened, have you generally found that you overetimated or underestimated its impoertance or you saw things in the right proportion?
12. How often do you have the feeling that there's little meaning in the things you do in your daily life?
13. How often do you have feelings that you're not sure you can keep under control?

The second factor consists of three items: 1, 7, and 10 (Cronbach's α = 0.60). The items are as follows:

1. Do you have the feeling that you don't really care about what goes on around you?
7. Doing the things you do every day is a source of deep pleasure and satisfaction or a source of pain and boredom?
10. Many people – even those with a strong character – sometimes feel like sad sacks (losers) in certain situations. How often have you felt this way in the past?

The last factor consists of five items: 2, 3, 4, 5, and 8 (Cronbach's α = 0.23). The items are as follows:

2. Has it happened in the past that you were surprised by the behavior of people whom you thought you know well?
3. Has it happened that people whom you counted on disappointed you?
4. Until now your life has had no clear goal or purpose at all or a very clear goal and purpose?
5. Do you have the feeling that you are being treated unfairly?
8. Do you have very mixed up feelings and ideas?

8.1.6 Brief COPE

The 28 items of the Brief COPE can be divided into 14 subscales, each comprised of two items, according to Carver (1997) (see 7.4.3). However, Carver suggests that each researcher should perform his or her own factor analysis and find higher-order factors, as each sample is different. Moreover, it makes no sense to use the total coping score for statistical analysis, as there is no meaningful overall score of coping. That is why it is more useful to find the relationships between the higher-order factors and other relevant variables.

The Kaiser-Meyer-Olkin test of sampling adequacy showed a value of 0.91, which is considered 'marvelous,' according to Kaiser (1974) and Bartlett's test of sphericity was significant (see Table 7). Therefore, the author proceeded to perform the PAF factor analysis.

Criteria for Extracting the Factors: the Eigenvalues, Percent of Variance Extracted, and the Scree Plot

Five factors had eigenvalues greater than 1.00, but only three explained more than 5 percent of the variance. The scree plot suggested a two-factor solution (see Figure 7). Thus, the author opted for a two-factor solution, as this also makes most sense intuitively (see Figure 8). The factors are: approach coping (Cronbach's alpha = 0.91) and avoidance coping (Cronbach's alpha = 0.80). The first factor explained 37.3 percent of the variance and the second factor 7.8 percent. Both factors together explain 45.1 percent of the variance.

Two items of the scale regarding alcohol and substance abuse were excluded from the analysis, as few people were willing to answer these items. Other authors such as Lawrence and Fauerbach (2003) using the COPE scale also excluded these coping strategies due to the same problems. Approach coping corresponds to both of Lazarus and Folkman's (1980) coping strategies, i.e. "problem-focused coping" and "emotion-focused coping," whereas avoidance coping seems to be a completely different construct.

Descriptions and Interpretations of the Factors of the Brief COPE Scale

The first subscale was called approach coping and comprises of the 15 items 2, 5, 7, 9, 10, 12, 14, 15, 17, 20, 22, 23, 24, 25, and 27 (Cronbach's $\alpha = 0.91$). The items are as follows:

2. I`ve been concentrating my efforts on doing something about the situation I`m in.
5. I've been getting emotional support from others.

7. I've been taking action to try to make the situation better.
9. I've been saying things to let my unpleasant feelings escape.
10. I've been getting help and advice from other people.
12. I've been using alcohol or other drugs to help me get through it.
14. I've been trying to come up with a strategy about what to do.
15. I've been getting comfort and understanding from someone.
17. I've been looking for something good in what is happening.
20. I've been accepting the reality of the fact that it has happened.
22. I've been trying to find comfort in my religion or spiritual beliefs.
23. I've been trying to get advice or help from other people about what to do.
24. I've been learning to live with it.
25. I've been thinking hard about what steps to take.
27. I've been praying or meditating.

The second subscale was called avoidance coping and comprises of the 11 items 1, 3, 6, 8, 13, 16, 18, 19, 21, 26, and 28 (Cronbach's $\alpha = 0.80$). The items are as follows:

1. I've been turning to work or other activities to take my mind off things.
3. I've been saying to myself "this isn't real".
6. I've been giving up trying to deal with it.
8. I've been refusing to believe that it has happened.
13. I've been criticizing myself.
16. I've been giving up the attempt to cope.
18. I've been making jokes about it.
19. I've been doing something to think about it less, such as going to movies, watching TV, reading, daydreaming, sleeping, or shopping.
21. I've been expressing my negative feelings.
26. I've been blaming myself for things that happened.
28. I've been making fun of the situation.

8.1.7 Social Support Scale

In the F-SOZU-22, perceived social support is assessed using three main scales: emotional support, tangible support, and social integration. An additional scale measures perceived social strain. In a principle component factor analysis with Varimax rotation, Fydrich et al. (1987) found the postulated four-factor structure.

The Kaiser-Meyer-Olkin measure of sampling showed a value of 0.94, which means "marvelous," according to Kaiser (1974), and the Bartlett's test of sphericity was significant, showing that a factor analysis with this data was feasible (see Table 8).

Criteria for Extracting the Factors for the Social Support Scale: the Eigenvalues, Percent of Variance Extracted, and the Scree Plot

Three factors have eigenvalues larger than 1.00 and together explain 54.4 percent of the variance. Although the scree plot suggests a two-factor solution (see Figure 9), the three-factor solution was chosen to maximize the explained variance (see Figure 10). The factors are 'emotional support' (Cronbach's alpha = 0.83), 'social embeddedness' (Cronbach's alpha = 0.84), and 'isolation' (Cronbach's alpha = 0.63). The first factor explains 41.0 percent of the variance, the second 8.7 percent, and the third 4.7 percent.

Descriptions and interpretations of the factors of the Social Support Scale SOZU-K-22

The first subscale was called "emotional support" and comprises of the 12 items 3, 7, 10, 11, 13, 14, 15, 16, 19, 20, 21, and 22 (Cronbach's $\alpha = 0.83$). The items are as follows:

3. It is important for my friends/acquaintances to hear my opinion about certain subjects.
7. I have friends/family members who are good listeners, if I want to talk about important personal matters.
10. When I am ill, I can ask friends/family members without hesitation to run important errands (e.g. shopping) for me.
11. If I feel greatly depressed, I know to whom I can turn.
13. There are people who share sorrow and joy with me.
14. With some friends/family members I can be very cheerful.
15. I know one person very well in whose presence I experience a great sense of wellbeing.
16. I know enough people who would really help me if I didn't know what to do.
19. I know enough people with whom I have a really good relationship.
20. There is a community of people (family, friends), to whom I feel I belong.
21. Through my circle of friends/acquaintances I often receive good practical advice (e.g. a good doctor) and other important information.
22. There are people to whom I can show all of my emotions without being embarrassed.

The second subscale was called "social embeddedness" and comprises of the six items 1, 2, 5, 6, 9, and 17 (Cronbach's $\alpha = 0.84$). The items are as follows:

1. I know people who would take care of my flat (plants, pets) when I'm not home.
2. There are people who accept me as I am.
5. I know one person very well whom I can expect to help me in any situation.
6. If I have the need, I can borrow tools or food.
9. I have friends/family members who would spontaneously give me a hug.
17. There are people who stick by me, even when I make mistakes.

The third subscale was called "isolation" and comprises of the four items 4, 8, 12, and 18 (Cronbach's $\alpha = 0.63$). The items are as follows:

4. I wish other people would show more understanding and affection towards me.
8. I hardly know anyone, with whom I would want to spend my free time.
12. I often feel like an outsider.
18. I wish I had more security and closeness.

8.2 Between-Group Comparisons

In order to verify the first and second hypotheses (*Tsunami victims are psychologically more psychologically affected in the long term than non-tsunami victims)*, the means of the sum total of the test scores of the control group (not affected by the tsunami) and the treatment group (affected by the tsunami) were compared to determine whether there was a significant difference. Whether a village was affected by the tsunami or not was the independent variable. The dependent variables were the mean test scores of the questionnaires.

As seen in Table 4, the tsunami–affected group shows higher overall scores on both the IES and BSI. The differences are significant with $p<.001$ for the IES and $p<.05$ for the BSI scores.

Table 4

Comparison of Means between the Affected and not Affected Group Regarding IES and BSI Scores

		IES Score	BSI Score
Not affected	Mean	21.96	38.42
	SD	17.77	37.95
Affected	Mean	38.73	49.02
	SD	16.61	41.26
Total	Mean	31.44	44.22
	SD	19.02	40.09

Table 5 shows the group statistics for the two t-tests performed to test whether there are significant differences between the affected and not affected group, one t-test for the IES-R and one t-test for the BSI. Table 6 shows the t values, significance levels, and confidence intervals.

Table 5

Group Statistics for t-Tests

	Affected by the tsunami	N	Mean	SD	Std. Error Mean
Sum IES-R	Not affected	141	21.96	17.77	1.50
	Affected	183	38.73	16.61	1.23
Sum BSI	Not affected	153	38.42	37.95	3.07
	Affected	185	49.02	41.26	3.03

Table 6

Independent Samples t-Tests for IES-R and BSI Mean Scores

	Levene's Test for Equality of Variances		t-Test for Equality of Means				
	F	Sig.	t	df	Sig. (2-tailed)	95% confidence interval Upper	Lower
Sum IES-R							
Equal variances assumed	0.37	0.55	-8.74	322	p < .001	-20.54	-12.99
Equal variances not assumed			-8.66	290	p < 0.001	-20.58	-12.96
Sum BSI							
Equal Variances Assumed	0.35	0.55	-2.44	336	p = 0.015	-19.15	-2.04
Equal Variances not assumed			-2.46	332	p = 0.015	-19.08	-2.11

8.2.1 Effect Size and Test Power

The effect size of the t-test for the IES was d = 0.98, which is considered large (0.20 is considered small, 0.50 medium, and 0.80 large, according to Bortz und Döring, 2006, p.606). The test power is 1-β = 1.00, which represents a large effect (1-β = .20-.50 represents a small effect, 1-β= .50-.80 a medium effect, and 1-β ≥ .80 a large effect). With missing values excluded, the sample size of the not affected group was 141, and the sample size of the affected group was 183. The non-centrality parameter δ was 8.70 and the df 322.

The effect size for the t-test of the BSI d = 0.27 is considered small, according to Bortz and Döring (2006, p.606). The test power is 1-β = 0.45, which is considered a small effect. With missing values excluded, the sample size of the not affected group was 155, and the sample size of the affected group 185. The non-centrality parameter δ was 2.46 and the df 338.

Subscale Comparisons

Table 7 shows the comparison of the mean subscale test scores between the affected and not affected group.

Table 7

Comparison of Means between the Affected and not Affected Group Regarding the Subscales of the IES-R and BSI

		IES		BSI		
		Intrusion/ Arousal	**Avoidance**	**Anxiety/ Depression/ Somatization**	**Hostility**	**Interper-sonal Sensitivity**
not affected	Mean	15.37	8.76	23.24	8.20	8.51
	SD	13.15	7.21	22.04	9.50	8.76
affected 10.23	Mean	25.95	14.21	33.50	8.53	
	SD	11.49	6.58	26.98	8.46	8.81
total	Mean	21.38	11.87	29.03	8.39	9.48
	SD	13.29	7.36	25.43	8.92	8.82
Effect size d		0.85	0.79	0.42	0.04	0.20
Test power 1-β		1.00	0.99	0.92	0.02	0.25

The differences between the means were significant with $p < 0.01$ for the 'intrusion' and 'avoidance' scores of the IES-R and the 'anxiety/depression/somatization' score of the BSI. The differences between the means for the 'hostility' and 'interpersonal sensitivity' scores of the BSI were not significant, with $p = 0.71$ and $p = 0.06$, respectively. The effect sizes were largest for the 'intrusion' and 'avoidance' subscales of the IES-R. The measure of association between being affected by the tsunami and the test scores was also highest for the 'intrusion' and 'avoidance' subscales of the IES-R (eta squared = 0.16 and 0.14, respectively, meaning that being affected by the tsunami explained 15.6 percen and 13.5 percent of the variance). For the 'anxiety/depression/somatization' subscale of the BSI, the association was much lower (eta squared = 0.04), and for the other subscales of the BSI, there was virtually no association.

8.2.2 Prevalence Rate of PTSD in the Affected Group

If a cut-off score of 24 was used to discriminate participants suffering from clinical PTSD and those who are not, as was done with a Japanese sample by Inoue et al. (2006), in the group affected by the tsunami, 74.9 percent were suffering from PTSD at the time of data collection. In the not affected group, only 35.5 percent were suffering from clinical PTSD. If both samples are pooled together, 57 percent of the total group were suffering from PTSD. The chi-square test is significant at the .001 level. The author chose the cut-off score used in a Japanese sample, as both the Japanese and the Indian culture are collectivistic cultures. In this way, we can expect similarities in the manifestation of symptoms.

8.3 Correlations

The third and fourth hypotheses (*Protective factors such as a sense of coherence and perceived social support decrease the extent of traumatization*) were tested on the tsunami-affected group using only Pearson's product moment correlations. This method highlights the relationship between psychological distress and protective factors. Table 8 shows the correlation matrix.

Table 8

Correlation Matrix for the IES, BSI, Social Support, SOC, and Coping Scores

	IES score	IES Intrusion	IES Avoidance	BSI score	Social Support	SOC
IES score	1					
IES Intrusion	.950**	1				
IES Avoidance	.834**	.620**	1			
BSI score	.605**	.644**	.334**	1		
Social Support	**-.164***	-.121	-.108	**-.239****	1	
SOC	-.140	**-.290****	.132	**-.591****	-.028	1

Note: ** *Correlation is significant at the 0.01 level (2-tailed).* * *Correlation is significant at the 0.05 level (2-tailed).*

In the tsunami-affected group, a significant negative correlation was found between SOC and BSI ($r = -.59$; $p < 0.01$). SOC is also negatively correlated with IES ($r = -.14$; $p > 0.05$), but not significantly. However, SOC is significantly negatively correlated with the intrusion subscale of the IES ($r = -.29$; p <

0.01). Social support is also negatively correlated both with IES ($r = -.16$; $p < 0.05$) and BSI ($r = -.24$; $p < 0.01$), which means that the higher the perceived social support, the less the extent of traumatization or any other psychiatric symptoms. However, social support is not correlated with SOC.

Table 9 shows the correlations between approach coping, avoidance coping, and the other variables in the tsunami-affected group.

Table 9

Correlation Matrix for the Approach Coping, Avoidance Coping, SOC, Social Support, IES-R, and BSI Scores

	IES-R	BSI	SOC	Social supp.	Approach Coping	Avoidance coping
IES	1					
BSI	.605**	1				
SOC	-.140	-.591**	1			
Social supp.	-.164*	-.239**	-.028	1		
Appr. Coping	**.319****	**.324****	.111	-.074	1	
Avoid. Coping	**.214****	**.335****	**-.267****	**-.260****	**.450****	1

*Note: ** Correlation is significant at the 0.01 level (2-tailed). * Correlation is significant at the 0.05 level (2-tailed).*

Significant negative correlations were found between avoidance coping and SOC ($r = -.27$; $p < 0.01$), as well as between avoidance coping and social support ($r = -.26$; $p < 0.01$). No significant correlations were found between approach coping and SOC, or between approach coping and social support. These findings are not in favor of hypothesis 7, which states that SOC should be positively correlated with effective coping strategies (in this sample, avoidance coping (see 8.4). Both coping styles are significantly positively correlated with IES-R as well as BSI scores ($p < 0.01$).

8.4 Regression Analyses

A regression analysis with the tsunami-affected group showed that perceived social support negatively predicts both IES-R and BSI (standardized beta = -0.13; R square = 0.02; $p = 0.55$ and standardized beta = -0.19; R square = 0.04; $p< 0.01$, respectively). Moreover, a strong sense of coherence predicted a low score on the BSI, with the standardized beta coefficient being -0.52 (R square =

0.27; p < 0.01) When SOC predicts IES-R, the standardized beta coefficient is -0.10, but not significant (R square = 0.10; p = 0.14). These findings also support hypotheses 3 and 4.

In the tsunami-affected group, regression analyses with approach and avoidance coping predicting IES-R and BSI scores, led to the following results: Approach coping significantly predicted a higher IES-R score (standardized beta = 0.28; R square = 0.10; p < 0.01), whereas avoidance coping did not significantly predict a lower IES-R score (standardized beta = 0.09; R square = 0.05; p > 0.05). Both approach coping and avoidance coping significantly predicted a higher BSI score (standardized beta = 0.22; R square = 0.11; p < 0.01 and standardized beta = 0.24; R square = 0.11; p < 0.01, respectively). In the total group, approach coping significantly predicted higher IES-R scores with beta being 0.30 (R square = 0.02; p < 0.01) and avoidance coping significantly predicted lower IES-R scores with beta being -0.23 (R square = 0.001; p < 0.01), which supports hypothesis 5 (*Avoidance coping is more effective in reducing PTSD than approach coping, as this study was conducted in a collectivistic cultural setting*). For the BSI, approach coping also significantly predicted higher scores with the standardized beta being 0.17 (R square = 0.06; p < 0.01). However, avoidance coping significantly predicted higher BSI scores (standardized beta = 0.22; R square = 0.05; p < 01).

In a regression analysis (with the total group) with being affected by the tsunami as the predictor and coping as the dependent variable, it was shown that being affected by the tsunami does not lead to more coping behavior. Being affected by the tsunami predicting approach coping is not significant (p = 0.54) and the beta coefficient would have only been -0.10 (R square = 0.01). When being affected by the tsunami predicted avoidance coping, this prediction was significant (p< 0.01) and the standardized beta coefficient was -0.26 (R square = 0.07). This means, that being affected by the tsunami leads to less coping behavior, especially avoidance coping. These findings do not support hypothesis 6 (*More coping efforts took place after the tsunami, i.e. the tsunami victims scored higher on all coping factors than the non-affected group*).

When performing a regression analysis (with the affected group) to determine whether SOC would predict more effective coping strategies (here: avoidance coping), a negative prediction was found (standardized beta = -0.34; R square = 0.11; p <0.01). The higher participants scored on the SOC scale, the lower was their reported avoidance coping behavior. In this way, hypothesis 7 (*SOC is positively correlated with more effective coping strategies and serves as a predictor for these coping strategies)* was not confirmed, if one considers avoidance coping the most effective coping behavior in a collectivistic setting. SOC was also a negative predictor for approach coping (standardized beta = -

0.20; R square = 0.04; $p < .001$), so it is evident that SOC definitely does not predict more coping behavior.

Stepwise Hierarchical Regression Analyses

Finally, stepwise hierarchical multiple regressions with the tsunami-affected sample showed which variables contributed the most to the general variance. In a first stepwise multiple regression model (see Table 10), approach coping, avoidance coping, SOC, and social support were entered into the stepwise regression model for the IES-R score. Only approach coping was significant in predicting the IES-R score (standardized beta = 0.32; $p<.001$), with R square being 0.10. This means that approach coping predicted 10. percent of the variance of IES-R scale score.

Table 10

Stepwise Multiple Regression Model 1 for the Tsunami-Affected Sample with IES-R as Dependent Variable

Variable	Unstd. B	Std. Error	Std. β	t	Sig.
Approach Coping	0.57	0.11	0.32	3.02	$p < 0.001$

In a second stepwise multiple regression model, the variables SOC, approach coping, and social support were found to be significant in predicting the BSI score. SOC was entered in the first step, SOC and approach coping in the second step, and SOC, approach coping, and social support in the third step. The R square changes were 0.27, 0.07, and 0.04, respectively. The standardized beta coefficients were -.49 for SOC, .26 for approach coping, and -.19 for social support. All three beta coefficients were significant with $p < .001$ (see Table 11). A strong sense of coherence predicted a low score on the BSI, whereas approach coping increased the extent of psychological distress. Social support again decreased the score on the BSI. The R square was .37, which means that SOC, approach coping, and social support together predicted 31.9 percent of the variance of the BSI score.

Table 11

Stepwise Multiple Regression Model 2 for the Tsunami-Affected Sample with BSI as Dependent Variable

Variable	Unstd. B	Std. Error	Std. β	t	Sig.	R square change
Step 1						0.266
SOC	-1.48	0.16	-0.52	13.98	$p < 0.01$	
Step 2						0.072
SOC	-1.40	0.16	-0.49	-9.01	$p < 0.01$	
Approach Coping	1.19	0.24	0.27	5.02	$p < 0.01$	
Step 3						0.035
SOC	-1.41	0.15	-0.49	-9.36	$p < 0.01$	
Approach Coping	1.13	0.23	0.26	4.85	$p < 0.01$	
Social Support	-0.42	0.12	-0.19	-3.57	$p < 0.01$	

When multiple regression analyses were performed with the total sample, the results slightly differed. Approach coping, avoidance coping, SOC, and social support were entered into the analysis predicting the IES-R score. In the first step, the variable social support was entered, in the second step social support and approach coping, and in the third step social support, approach coping, and avoidance coping. The R square changes for each step were 0.034, 0.020, and 0.037, respectively. The R square was 0.091 for the third step, meaning that the variables social support, approach coping, and avoidance coping explained 9.1 percent of the variance for the IES score. The standardized beta coefficients were -0.21 for social support, 0.33 for approach coping, and -0.27 for avoidance coping. For all three beta coefficients, p was smaller than 0.001 (see Table 12).

Table 12

Stepwise Multiple Regression Model 1 for the Total Sample with IES-R as Dependent Variable

Variable	Unstd. B	Std. Error	Std. β	t	Sig.	R square change
Step 1						0.034
Social Support	-0.22	0.06	-0.19	-3.79	p < 0.01	
Step 2						0.020
Social Support	-0.22	0.06	-0.19	-3.83	p < 0.01	
Approach Coping	0.24	0.08	0.14	2.94	p < 0.01	
Step 3						0.037
Social Support	-0.25	0.06	-0.21	-4.40	p < 0.01	
Approach Coping	0.56	0.11	0.33	4.97	p < 0.01	
Avoidance Coping	-0.65	0.16	-0.27	-4.05	p < 0.01	

When approach coping, avoidance coping, social support, and SOC were entered into a stepwise multiple regression analysis to predict the BSI score, SOC was entered in the first step, SOC and social support in the second step, and SOC, social support, and approach coping in the third step. R square changes were 0.186, 0.040, and .0250, respectively. In the third step, the predictor variables explained 25.1 percent of the variance of the BSI score (R square = 0.251). The standardized beta coefficients were -0.40 for SOC, -0.20 for social support, and 0.16 for approach coping. The beta coefficients were significant with $p < .001$ (see Table 13).

Table 13

Stepwise Multiple Regression Model 2 for the Total Sample with BSI as Dependent Variable

Variable	Unstd. B	Std. Error	Std. β	t	Sig.	R square change
Step 1						0.186
SOC	-1.23	0.13	-0.43	-9.62	p < 0.01	
Step 2						0.040
SOC	-1.23	0.13	-0.43	-9.83	p < 0.01	
Social Support	-0.52	0.11	-0.20	-4.59	p < 0.01	
Step 3						0.025
SOC	-1.14	0.16	-0.40	-9.05	p < 0.01	
Social Support	-0.52	0.11	-0.20	-4.68	p < 0.01	
Approach Coping	-0.59	0.16	0.16	3.64	p < 0.01	

8.5 Comparing Demographic Factors in Relation to PTSD

In addition to the psychometric scales, the subjects were given a demographic sheet to assess the severity of their personal trauma, for example, whether they lost a family member, their home, or their means of earning a livelihood. The villagers were also questioned about their level of education, age, marital status and other demographic data. In this way, demographic data, as well as the severity of trauma, could be included in the regression analysis in order to provide insight into why some people were more affected by the trauma than others.

According to hypothesis 8.1, the following demographic factors serve as protective factors, i.e. they should be negatively associated with PTSD and other psychological distress and predict a lower level of symptoms: high education, male gender, being married, middle age, and having received a new house and/or a new job.

Hypothesis 8.2 postulates that the following demographic factors serve as risk factors, i.e. they are positively associated with PTSD and other psychological distress and predict a higher level of symptoms: low education, female gender, widowhood, young or old age, or chronic illness.

Regarding the overall score of the IES, i.e. the level of PTSD symptoms, it is evident that marital status, profession, and age group play a significant role (see Table 14). Married people are less traumatized, as well as people who have a job other than fishing. The people in the youngest age group are more traumatized than the other age groups, which correspond to students in college being more traumatized.

The eta coefficient gives information about the strength of association between a nominal or ordinal independent and an interval dependent variable. It can only vary between 0 and 1. Eta squared indicates how much variance of the dependent variable is explained by the independent variable (Bryman & Cramer, 2009).

Table 14

Demographic Factors and IES-R Scores

Demographic Factors		Means of the IES-R Score	Level of Sig. for Mean Differences	Eta	Eta square
Marital Status	Single	45.56	$p = .02$	0.22	0.05
	Married	**36.89**			
	Widowed	44.52			
Profession	Housewife/ No job	39.67	$p = .01$	0.25	0.06
	Fisher	38.05			
	Other job	**25.70**			
	In college	**51.29**			
Age Groups	up to 25	**50.83**	$p = .05$	0.21	0.04
	26-35	39.47			
	36-60	36.85			
	over 60	38.82			

When reviewing at the overall score of the BSI, i.e. general psychological distress, it becomes clear that marital status, chronic illness, and age group play a significant role (see Table 15). Widowed people are the most psychologically distressed, as well as people who have some form of chronic illness. Moreover, the very young and very old are more distressed.

Table 15

Demographic Factors and BSI Scores

Demographic Factors		Means of the BSI Score	Level of Sig. for Mean Differences	Eta	Eta square
Marital Status	Single	39.80	$p = .01$	0.33	0.11
	Married	45.39			
	Widowed	**79.90**			
Chronic Illness	Yes	**69.14**	$p < .01$	0.27	0.08
	No	42.51			
Age Groups	up to 25	**60.25**	$p < .01$	0.24	0.06
	26-35	45.04			
	36-60	44.34			
	over 60	**74.81**			

In this way, both hypotheses 8.1 and 8.2 have been partly confirmed, as there is a negative association between the postulated protective factors and symptom severity and a positive association between the postulated risk factors and symptom severity. However, no gender differences or differences regarding education level were discovered.

Next, a multiple regression analysis was performed to investigate the predictive power of the demographic variables in explaining the variance of the IES-R and BSI scores in the tsunami-affected group. In a first model, age, gender, marital status, profession, education, and chronic illness were entered into the model to predict the IES score. However, none of the variables significantly predicted the IES score at $p < 0.05$. Furthermore, the model only explained 1.2 percent of the variance. The same variables were entered into a second model to predict the BSI score (see Table 16). Here, marital status significantly predicted the BSI score (standardized beta = 0.24; $p<.001$), as well as the existence of any chronic illness (standardized beta = 0.21; $p = .001$). R square changes were 0.064 and 0.43. Marital status and chronic illness together explained 10.8 percent of the variance. Therefore, hypotheses 8.1 and 8.2 were only partly confirmed, as there is an association between demographic factors and symptom severity, although the demographic variables themselves (except 'marital status', and 'chronic illness') do not predict symptom severity. The only demographic factor acting as a protective factor is being married.

Table 16

Stepwise Multiple Regression Model for the Demographic Variables Predictng BSI Scores

Variable	Unstd. B	Std. Error	Std. β	t	Sig.	R square change
Step 1						0.064
Marital Status	18.80	4.72	0.25	3.99	$p < 0.01$	
Step 2						0.043
Marital Status	17.40	4.63	0.24	3.75	$p < 0.01$	
Chronic Illness	17.71	5.31	0.21	3.34	$p = 0.01$	

Variables of Trauma Severity

In order to assess which variables of trauma severity contribute most to the variance of the IES-R and BSI scores in the tsunami-affected sample, the following variables were entered into a multiple regression model: loss of family or friends, loss of home, loss of job, and personal injury during the tsunami. In the model predicting the IES-R score, none of the variables significantly predicted the IES-R scores and the model only explained 1.3 percent of the variance. In the model predicting the BSI score, however, none of the variables significantly predicted the BSI score, either. In addition, the model only explained 0.6 percent of the variance.

Similar results were found for the total sample. Only the variable 'personal injury' significantly predicted higher IES-R scores (standardized beta = 0.28; R square = 0.075; p<.001). None of the variables significantly predicted the BSI score (R square = 0.008; all of the p values > 0.05). In this way, hypothesis 9 was only confirmed as far as 'personal injury/disabled' was concerned, but otherwise not (*Factors determining the severity of the trauma, such as loss of family members or friends, loss of home, damage of home, loss of job, or personal injury, predict a higher level of PTSD symptoms and other symptoms of psychological distress*).

8.6 Path Analysis Model

In order to explore the causal relationships between several variables at once in a more elaborate way, the author performed a path analysis with the data.

8.6.1 The Hypothesized Model

The hypothesized model is presented in Figure 1 The author postulated in hypothesis 10 that social support would have a direct effect on PTSD and general psychological distress. The effect of SOC on PTSD and general psychological distress should be both directly and indirectly mediated by coping behavior, i.e. approach and avoidance coping. Whether people are affected by the tsunami or not should have a direct effect on coping behavior and social support, as well as on both variables of traumatic stress. Moreover, approach coping should positively predict PTSD and general psychological distress, whereas avoidance coping should negatively predict variables of traumatic stress.

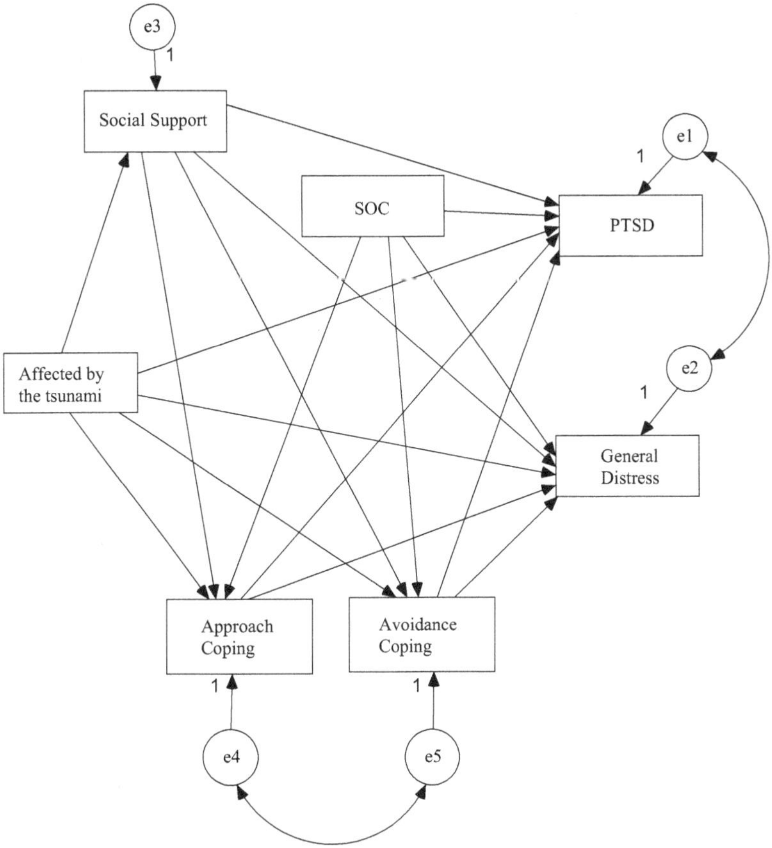

Figure 1 The hypothesized model of the relationship between protective factors and traumatic stress symptoms.

8.6.2 Data Analysis

In order to test the hypothesized model, structural equation modeling (SEM) was performed. SEM is a useful statistical technique that enables researchers to statistically test the fit of a model of the relationship between the variables in their study (Byrne, 2010). In this study, a path analysis was observed with observed variables. After evaluating the model fit to determine how adequately the model accounts for the data, the model fit was improved by respecifying the model. To guide the respecification process, modification indices were used. After determining the best fitting model, each individual path in the model was tested by means of standardized regression coefficients.

In SEM, it is standard to use multiple criteria for evaluating model fit. In the literature, numerous goodness-of-fit indices can be found, for example: Pearson chi-squared divided by the degrees of freedom (χ^2 / *df;* criterion: < 3.0), comparative fit index (CFI; criterion: >0 .95), goodness-of-fit index (GFI; criterion: > 0.9), Tucker-Lewis Index (TLI; criterion: > 0.95), root-mean-square error of approximation (RMSEA; criterion: < 0.08), Akaike's information criterion (AIC; criterion: low), and expected cross-validation index (ECVI; criterion: low) (Byrne, 2010).

8.6.3 Results of Path Analysis

Data Screening

Missing data were substituted using the series mean. No evidence of multicollinearity was found – correlations between variables were modest and all multiple correlations (R square) were well below 0.85, a cut-off score suggested by Kline (2005). Outliers were detected using stem and leaf displays and box plots. All variables were somewhat skewed. IES was slightly positively skewed (skewness = 0.27, kurtosis = 0.08) and BSI positively skewed (skewness = 1.44, kurtosis = 2.54). Social support was negatively skewed (skewness = -.085, kurtosis = 2.48) and SOC slightly positively skewed (skewness = 0.21, kurtosis = 0.09). Approach coping was slightly negatively skewed (skewness = -0.38, kurtosis = 0.24) and avoidance coping slightly positively skewed (skewness = 0.52, kurtosis = 0.24). When taking the sum of the COPE scale into account, the skewness is -0.18 and the kurtosis -0.05. However, the distribution was within the accepted range of univariate normal distributions according to the cut-off guidelines suggested by Kline (2005) (skew < 3 and kurtosis < 10).

Testing the Hypothesized Model

Before arriving at the hypothesized model, a preliminary path analysis model with latent variables and the subscales of all measures as observed variables was tested. However, a model with latent variables was not feasible, as the subscales of the measures apparently do not adequately represent the latent constructs. It was not possible to postulate a latent "coping" variable. This is why the author postulated a hypothesized model with only observed variables. Moreover, it is more meaningful to assess the relationship between the different coping strategies and the variables of traumatic stress than to use a latent general coping variable of "coping efforts."

However, Figure 1, which shows the hypothesized model, would have only fit the data if all error terms had been correlated with each other. Therefore, the residual error explains more variance than the variables themselves. There must be a large amount of systematic error in the scales. Nevertheless, when the sum scores of the scales as well as the sum scores of the coping subscales were used as observed variables, the model fit the data quite well. Additionally, the variable "Were you affected by the tsunami?" was added in order to test the effect of exposure to the tsunami. The new model is shown in Figure 2.

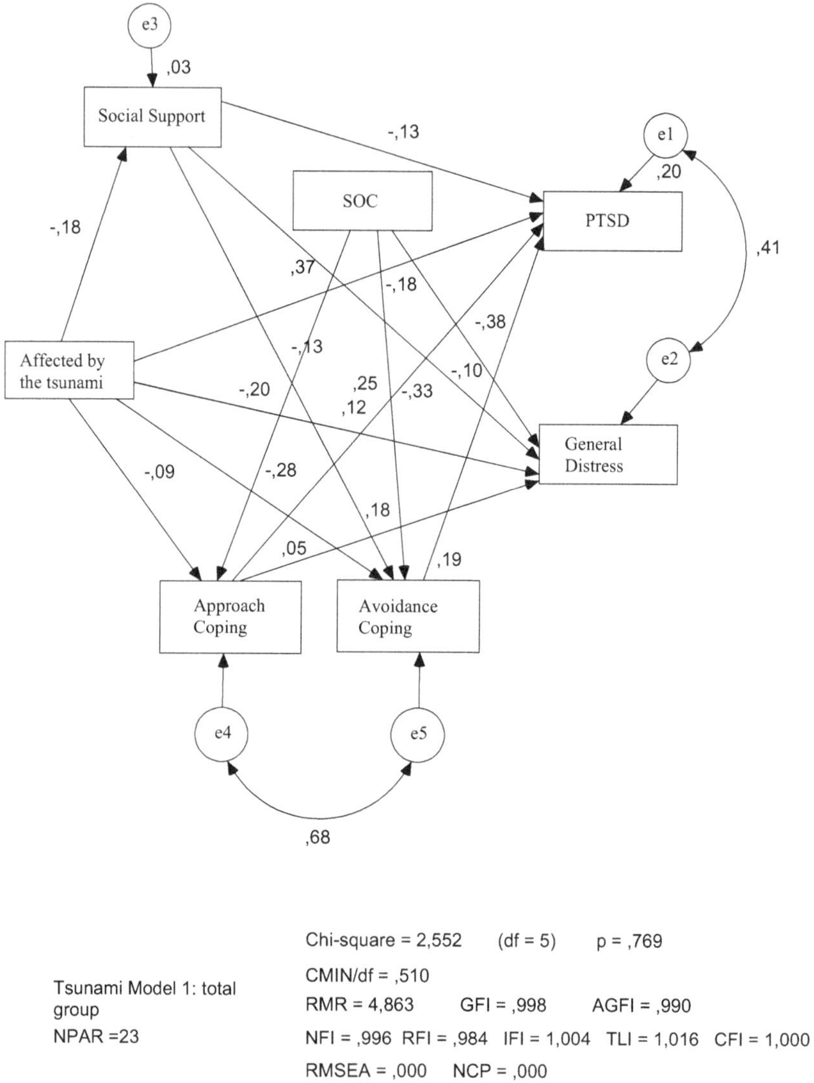

Figure 2. A first path analysis model for the total group.

This model, led to a χ^2 of 2.55 and a p value of 0.77. In a further step, the pathway between SOC and PTSD, as well as the pathway between avoidance coping and the general psychological distress were deleted, as these regressions were not significant. In addition, the pathway between social support and approach coping was deleted, as this regression was also not significant. According to the theoretical background, in collectivistic cultures, approach coping leads to higher levels and avoidance coping to lower levels of traumatic stress. These predictions were confirmed by the model, with the exception of the association between avoidance coping and general psychological distress, which was not significant (approach coping predicting PTSD: beta = 0.25; approach coping predicting general psychological distress: beta = 0.18; avoidance coping predicting PTSD: beta = -.10). The assumptions of social support and SOC predicting lower levels of traumatic stress (social support predicting PTSD: beta = -0.13; social support predicting psychological distress: beta = -0.18; SOC predicting psychological distress: beta = -0.38) and being affected by the tsunami leading to higher levels of traumatic stress (affected predicting PTSD; beta = 0.37; and affected predicting psychological distress: beta = 0.12) were confirmed by the model. Social support predicted lower levels of avoidance coping (beta = -0.13). SOC predicted lower levels of coping efforts (beta = -0.20 for approach and -0.32 for avoidance coping), which is a contradictory finding. Moreover, being affected by the tsunami predicted lower levels of social support (beta = -0.18) as well as lower levels of approach (beta = -0.09) and avoidance (beta = -0.28) coping.

In a further analysis, two separate path diagrams were constructed for the tsunami-affected group and the non-affected group to investigate group differences. Figure 3 shows the path diagram for the tsunami-affected group.

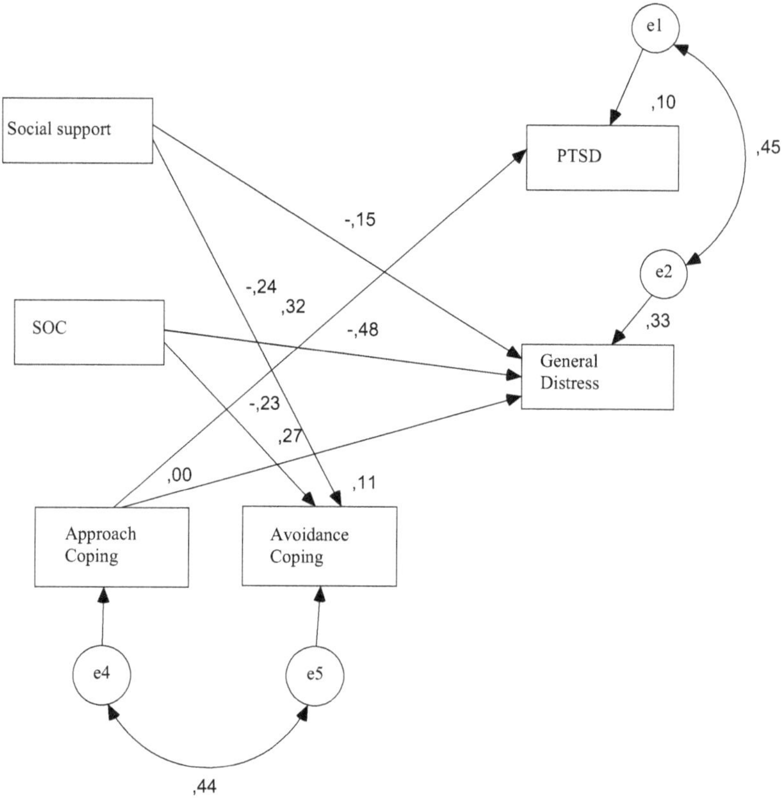

Chi-square = 9,157 (df = 7) p = ,242

Tsunami Model 2: Affected Group NPAR = 14

CMIN/df = 1,308

RMR = 23,651 GFI = ,987 AGFI = ,962

NFI = ,967 RFI = ,930 IFI = ,992 TLI = ,983 CFI = ,992

RMSEA = ,036 NCP = 2,157

Figure 3. A second path analysis model for the tsunami-affected group.

This model, reveals that social support predicts lower levels of traumatic stress, but only regarding the general psychological distress (beta = -0.15); it does not significantly predict PTSD. However, social support still predicts lower levels of avoidance coping (beta = -0.24). SOC predicts lower levels of general psychological distress (beta -0.48), as well as lower levels of avoidance coping (-0.23). In this group, approach coping also predicts higher levels of PTSD (beta = 0.32) and general psychological distress (beta = 0.27). Avoidance coping has no significant predicting quality. Figure 4 shows the model for the non-affected group.

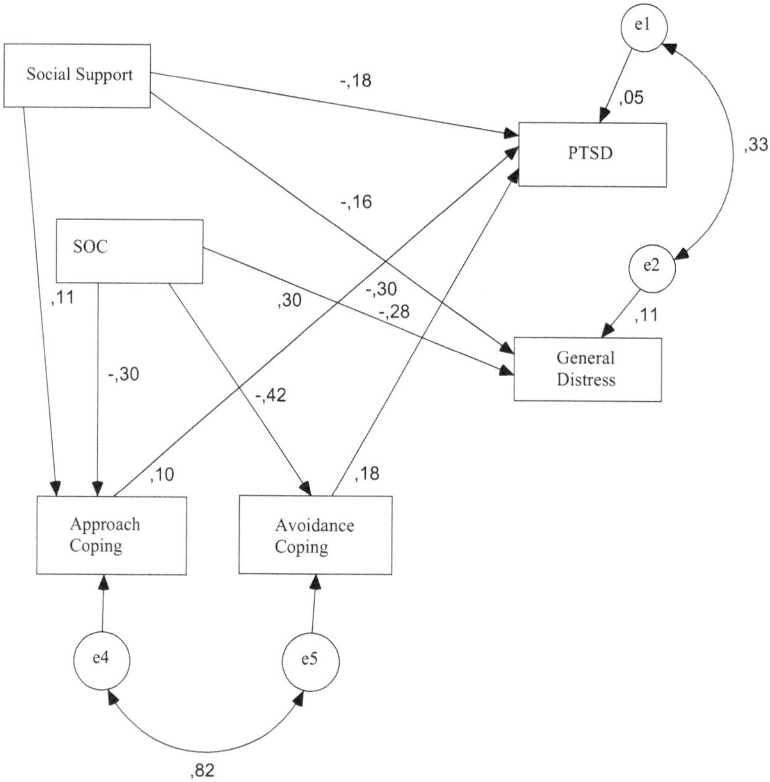

Tsunami Model 3: non-affected group

NPAR = 16

Chi-square = 5,455 (df = 5) p = ,363
CMIN/df = 1,091
RMR = 15,465 GFI = ,990 AGFI = 957
NFI = ,982 RFI = ,947 IFI = ,998 TLI = ,995 CFI = ,998
RMSEA = ,023 NCP = ,455

Figure 4. A third path analysis model for the non-affected group.

In the non-affected group, it is evident that social support predicts lower levels of both PTSD (beta = -0.18) and general psychological distress (beta = -0.16). However, it also predicts higher levels of approach coping (beta = 0.11), which is contradictory to the above models. As in the other models, SOC predicts lower levels of general psychological distress (beta = -0.33) and lower levels of both approach (beta = -0.30) and avoidance coping (beta = -0.42). Approach coping predicts higher levels of PTSD (beta = 0.30) and avoidance coping lower scores (beta = -0.28). The general psychological distress score is not affected by either coping strategy.

9 Overview on the Confirmation of or Contradiction to the Hypotheses

The following chapter shall give an overview on whether or not and to what extent the hypotheses could be confirmed.

Hypothesis 1 "Tsunami victims show higher rates of PTSD (assessed with the Impact of Event Scale-Revised (IES-R) by Weiss and Marmar (1996)) than non-tsunami victims two and a half years after the tsunami" and 2 "Tsunami victims show higher rates of general psychological distress (assessed with the Brief Symptom Inventory (BSI) by Derogatis (1993)) than non-tsunami victims two and a half years after the tsunami" were confirmed, as the tsunami–affected group shows significantly higher overall scores on both the IES-R and BSI. In the IES-R, both "intrusion/arousal" and "avoidance" subscales showed significant differences, whereas in the BSI, only the subscale "anxiety/depression/somatization" was significantly higher in the tsunami-affected group. The other subscales "hostility" and "interpersonal sensitivity" did not show any significant differences between the two groups. Moreover, significant differences in prevalence rates existed between the two groups: in the tsunami-affected group, 74.9 percent, and in the not affected group, 35.5 percent were suffering from clinically significant PTSD.

Hypothesis 3 "There is a negative correlation between Antonovsky's Sense of Coherence (SOC) (Antonovsky, 1987) and the level of PTSD" was only partly confirmed. SOC was negatively correlated with the IES-R score, but not significantly. However, SOC was significantly negatively correlated with the "intrusion/arousal" subscale of the IES-R, but not the "avoidance" subscale. Furthermore, SOC was significantly negatively correlated with the BSI total score.

Hypothesis 4 "There is a negative correlation between perceived social support and the degree of PTSD" is confirmed, as perceived social support is significantly negatively correlated both with IES-R and BSI scores, which means that the higher the perceived social support, the less the extent of traumatization or other psychological distress. Both hypotheses 3 and 4 are also supported by regression analyses showing that SOC significantly predicts lower BSI scores and perceived social support significantly predicts both lower IES-R and BSI scores.

Hypothesis 5 "Avoidance coping is more effective in reducing PTSD than approach coping, as this study was conducted in a collectivistic cultural setting" was only partly confirmed in the tsunami-afftected group. Approach coping significantly predicted a higher IES-R score, whereas avoidance coping did not significantly predicted a lower IES-R score. Both approach and avoidance coping significantly predicted a higher BSI score. However, hypothesis 5 was verified for the total group. In the total group, approach coping significantly predicted a higher, and avoidance coping significantly predicted a lower IES-R score. For the BSI score, approach coping also significantly predicted higher scores, although avoidance coping did not significantly predict a lower BSI score.

Hypothesis 6 "More coping efforts took place after the tsunami, i.e. the tsunami victims scored higher on all coping factors than the non-affected group" was not confirmed. A regression analysis with being affected by the tsunami as the predictor and coping as the dependent variable showed that being affected by the tsunami does not lead to more coping behavior.

Hypothesis 7 "SOC is positively correlated with more effective coping strategies and serves as a predictor for these coping strategies" is also not confirmed. In this sample from a collectivistic culture, it is assumed that avoidance coping is the more efficient strategy. However, SOC does not predict more avoidance coping behavior. Rather, SOC significantly predicts lower levels of avoidance coping. Moreover, SOC significantly predicted less approach coping behavior, thus it is apparent that SOC does not predict more coping efforts in general.

Hypothesis 8.1 "The following demographic factors serve as protective factors, i.e. they are negatively correlated with PTSD and other psychological distress and predict a lower level of symptoms: high education, male gender, being married, middle age, or having received a new house and/or a new job" was partly confirmed. The variables being married, having some other occupation than fishing, and being in the middle- to old-agegroup were significantly associated with lower IES-R scores and thus serve as protective factors. The variables being married, having no chronic illness, and belonging to the middle-age group were associated with lower BSI scores. Hierarchical regression analyses revealed that marital status and having no chronic illness were protective factors. As opposed to the hypothesis, high education and male gender did not serve as protective factors. Having received a new house or job also did not influence the level of traumatization.

Hypothesis 8.2 "The following demographic factors serve as risk factors; i.e. they are positively correlated with PTSD and other psychological distress and predict a higher level of symptoms: low education, female gender, widowhood, young or old age, or chronic illness." Widowhood and young age were significantly associated with higher IES-R levels, and widowhood, chronic illness, and both young and old age with BSI scores. The level of education and female gender did not serve as risk factors. Thus, hypothesis 8.2 was also partly confirmed. The regression analysis indicated that being widowed and having some form of chronic illness served as risk factors.

Hypothesis 9 "Factors determining the severity of the trauma, such as loss of family members or friends, loss of home, damage of home, loss of job, or personal injury, predict a higher level of PTSD symptoms and other symptoms of psychological distress" was only confirmed as far as "personal injury" is concerned. The other variables of trauma severity did not predict higher IES-R or BSI scores in hierarchical regression analyses.

Hypothesis 10 "Social support has a direct effect on PTSD and other psychopathological symptoms. The effect of SOC on PTSD and other psychopathological symptoms is both direct and mediated by coping behavior, i.e. approach and avoidance coping. Whether people are affected by the tsunami or not has a direct effect on coping behavior and perceived social support, as well as on both variables of traumatic stress. Approach coping positively predicts PTSD and other psychopathological symptoms, whereas avoidance coping negatively predicts variables of traumatic stress. Perceived social support has an effect on coping behavior" was confirmed in a path analysis model.

According to the theoretical background, in collectivistic cultures, approach coping led to higher levels and avoidance coping to lower levels of traumatic stress. These predictions were confirmed by the model for the total group, with the exception of the association between avoidance coping and general psychological distress, which was not significant. The assumptions of social support and SOC predicting lower levels of traumatic stress and being affected by the tsunami leading to higher levels of traumatic stress were confirmed by the model. SOC predicted lower levels of both coping strategies, which is a contradictory finding. The assumption that the effect of SOC on PTSD and other psychopathological symptoms was mediated by coping behavior was not confirmed, as SOC did not predict lower levels of PTSD but only lower levels of psychological distress. SOC led to lower levels of both coping behaviors, but coping only predicted levels of PTSD, not psychological distress. Thus, the mediating effects were not confirmed. Being affected by the tsunami predicted less perceived social support and social support predicts less avoidance coping. Being affected by the tsunami predicted less approach coping behavior.

In the total group, social support predicted lower levels of avoidance coping. In the tsunami-affected group, social support still predicted lower levels of avoidance coping, but it did not significantly predict lower levels of PTSD, only general psychological distress. In this group, avoidance coping had no predicting quality. In the not affected group, however, social support predicted higher levels of approach coping.

10 Discussion

In conclusion, tsunami victims were still traumatized two and a half years after the tsunami and they showed greater psychological distress. In the tsunami-affected group, 44.8 percent of the people scored one standard deviation above the mean IES score of the total group. If the cut-off score of 24 is used for the IES, as was used in a Japanese sample (Inoue et al., 2006), 74.9 percent of the tsunami-affected group would be suffering from PTSD at the time of data collection. In the not affected group, only 35.5 percent were suffering from clinical PTSD. If both samples were pooled together, 57 percent of the total group would be suffering from PTSD, so it is evident that they were quite traumatized. When comparing the affected and not affected group, the affected group showed significantly more PTSD symptoms and general psychological distress symptoms.

Moreover, as hypothesized, SOC and social support decreased the level of traumatic stress. SOC significantly decreased the level of general psychological distress and intrusion symptoms of PTSD. Social support decreased both intrusion and avoidance symptoms of PTSD, as well as the level of general psychological distress.

The results of the study also indicated that avoidance coping is the coping strategy that is more effective in a collectivistic setting if the total sample is taken into account. Avoidance coping significantly decreased and approach coping significantly increased PTSD symptoms. Approach coping also significantly increased psychological distress. However, in the tsunami-affected sample, only approach coping significantly predicted a higher level of PTSD. Both approach and avoidance coping strategies were significantly positively correlated with IES-R as well as BSI scores ($p < 0.01$) and significantly predicted higher levels of general psychological distress. It is open to speculation that the manifestation of traumatic stress symptoms led to more coping efforts, hence the positive correlation. However, correlations do not give information about causality, and this last finding could also indicate that the more people suffer from symptoms of traumatic stress, the more they engage in coping efforts. Moreover, being affected by the tsunami is no significant predictor of higher coping scores, although both coping strategies are positively correlated with symptoms of traumatic stress. Thus, the event of the tsunami itself does not seem to trigger more coping responses, but the presence of PTSD symptoms does.

If one can determine the protective factors that mitigate the effects of traumatic stress, then it may be possible to use them as the focal point for the prevention of PTSD or other mental disorders or for intervention after a natural disaster. Therefore, in order to maintain the psychophysical health of individuals who have encountered a traumatic event, it is necessary to strengthen their sense of coherence and increase their sense of receiving social support.

10.1 SOC and Traumatic Stress

SOC significantly decreased the intrusion symptoms of PTSD and symptoms of general psychological distress. It is open to speculation why avoidance symptoms of PTSD were not decreased through a stronger SOC. Perhaps, avoidance symptoms actually help in dealing with the intrusion symptoms, as postulated by Antonovsky (1987). People oscillate between intrusion and avoidance until they can integrate the traumatic experience. In this way, avoidance behavior is actually an adaptive coping strategy, which is employed by people who have a strong SOC. Moreover, in this collectivistic culture sample, avoidance coping was found to be the more adaptive coping strategy.

In all three path analysis models, however, SOC did not significantly predict lower levels of PTSD, although it predicted lower levels of general psychological distress. The fact that SOC predicted lower levels of coping efforts might be interpreted in light of Lazarus and Folkman's (1984) theory of appraisal. According to Antonovsky (1987), a person with a strong SOC appraises a stressor not so much as harm, loss, or a threat, but more as a non-stressor (primary appraisal). Consequently, there is no need to engage in intense coping efforts.

10.2 Social Support and Traumatic Stress

Natural disasters have an effect both on the individual and on the community level of society. The response to the disaster by the social system depends on its inherent characteristics and dynamics. However, the cause, predictability, controllability, frequency of occurrence, scope of impact, and destructive power of the disaster also play a major role. Psychological disorders tend to be magnified if a large proportion of a community's resources are destroyed by the disaster, if the local environment is profoundly altered, or if permanent relocation is necessary. If there had been conflicts within the community beforehand, these may surface more strongly in times of crisis. Social control and participation can thus be weakened, traditional community groupings can be fragmented, the system of rights and obligations can break down, and racial or ideological discrimination in the distribution of relief aid can occur (Alexander, 1999).

However, during the aftermath, high value also tends to be given to community interests and low value to personal ones, and there generally is higher consensus on what is good, just, and correct. Altruism and heightened social support help individuals cope with the disaster and mitigate their stress level. This brief period after the disaster, usually a few weeks, is often referred to as a "therapeutic community" (Alexander, 1999) or "recovery environment" (Green, Wilson, & Lindy, 1985). People react to disasters either with activism of fatalism, the latter including supplication and prayer, but most likely manifesting itself as passive behavior (Alexander, 1999). In a study with older adults, Kaniasty, Norris, and Murrell (1990) investigated whether expected help after a flood was actually received. However, on both kin and non-kin help measures, most victims reported receiving little or no help at all. Thus, evidently their results do not correspond to those studies reporting widespread altruism following natural disasters.

In this dissertation study, perceived social support was investigated as opposed to actual received social support. Perceived social support significantly predicted lower levels of PTSD symptoms, as well as symptoms of general psychological distress. Thus, theories of social support decreasing symptoms of traumatic stress have been confirmed.

The assumption that social support predicts a lower level of traumatic stress has also been confirmed by the path analysis models with the exception of the tsunami-affected group. In this group, social support did not significantly predict lower levels of PTSD, only lower levels of general psychological distress. It can be speculated that in the face of devastating disaster, social support is not sufficient to compensate for the loss of life and property. However, if social support had been missing, the effects of the disaster would have been more dramatic (Solomon, et al. 1988). In the models with the total and the tsunami-affected group, social support predicted lower levels of avoidance coping, and in the non-affected group higher levels of approach coping. These findings suggest that social support enables people to actively change the situation or deal with their emotions, as they can access support from others as a coping strategy. However, trying to actively change the situation apparently did not reduce PTSD symptoms, since approach coping led to higher levels of PTSD symptoms.

10.3 Coping and PTSD

The regression analyses and path analysis models confirmed the assumption that in collectivistic cultures approach coping leads to higher levels and avoidance coping to lower levels of traumatic stress. However, in the tsunami-affected group, avoidance coping is not a significant predictor of traumatic stress. Seen from a different perspective, the tsunami-affected model suggests the reciprocal relationships between coping and PTSD symptoms. The fact that more coping efforts predict a higher level of PTSD symptoms can also be interpreted as higher levels of traumatic stress eliciting more coping efforts, especially more active, approach coping behaviors. Similar results were found in a study with Bhutanese torture victims living as refugees in Nepal by Emmelkamp, Komproe, van Ommeren, and Schagen (2002). These authors found a positive relationship between the number of coping strategies used and the level of anxiety and depressive symptoms. Furthermore, in a study with Palestinian and Israeli adolescents exposed to violent conflict, the participants in the more severe exposure group with the most PTSD symptoms reported using more adaptive as well as maladaptive coping strategies than those adolescents in the less severe exposure group. The authors performed a factor analysis on the Brief COPE scale and had found two factors which they named "adaptive" and "maladaptive" coping. Both factors were significant predictors for PTSD, even after controlling for exposure level and gender (Pat-Horenczyk et al., 2009). Finally, it is not certain whether participants affected by the tsunami would not be suffering even more from PTSD and other psychiatric symptoms if they had not undertaken intense coping efforts. Moreover, rather than actual coping, the attempts at coping were assessed; thus, both effective and ineffective strategies come into play.

In a study with victims of Hurricane Katrina in the United States, Glass et al. (2009) combined problem-focused and active emotion-focused coping into one problem-focused factor, which corresponds to this dissertation study's approach coping factor. These authors had similar findings in that their problem-focused coping also positively predicts PTSD. According to these authors, this may be due in part to the uncontrollability of such a natural disaster. People might have tried to use coping strategies that were useful in the past. However, in this scenario, the same coping efforts may have been unavailable or ineffective. This ineffective use of problem-focused coping may lead to more frustration and more psychological distress. Avoidant coping might be the more important coping strategy in explaining distress following an uncontrollable natural disaster (Glass et al., 2009). A study with parents of children with pediatric cancer in the United States revealed similar results (Greening & Stoppelbein, 2007). In the face of such an uncontrollable event, symptoms decreased when the parents used emotional regulation strategies and increased when they used problem-solving strategies. This result confirms the authors' hypothesized relation between problem management coping strategies and psychological symptoms when coping with an uncontrollable event. Thus, the higher efficacy of avoidance coping in decreasing symptoms of traumatic stress might be due to the uncontrollability of the event rather than specific cultural factors.

Moreover, in this dissertation study, the author found that affected individuals used less coping responses than individuals not affected by the tsunami.

The reason may be that exposure to a traumatic event leads to some kind of numbing reaction. However, if traumatic stress symptoms occur after the event, people start to engage in more coping efforts. In a study with victims of the Kashmir conflict in Indian Kashmir, participants directly exposed to the conflict also exhibited less coping behavior than an indirectly affected control group (Yaswi & Haque, 2008). According to these authors, social support plays an important role in effective coping, especially in a close-knit society such as the Kashmiri society. An individual's problem becomes the community's problem. In this way, the adverse effects of trauma are shared and thus decreased.

In this dissertation study, the author only assessed coping strategies typically employed in Western cultures with the Brief COPE scale and which could be grouped into the two factors 'approach' and 'avoidance' coping. This scale was developed from an individualistic cultural perspective. A coping measure developed from a collectivistic cultural perspective might have operated in a different way. Perhaps the COPE scale does not measure the full range of possible coping strategies in an Indian sample. In a qualitative study by Rajkumar, Premkumar, and Tharyan (2008), it was revealed that survivors of the 2004 tsunami in Tamil Nadu, South India, valued their ethno-cultural and community coping strategies, such as collectivizing personal trauma, reconstructing meaning following disaster using a fatalistic perspective, utilizing extended social supports, displaying grief and mourning publicly, and drawing on strongly rooted spiritual beliefs. They considered their survival as gift from God, and they did not wish to view themselves as victims needing external from outside. To begin, they never expected their life to be luxurious, and their lives as fisherfolk are inherently risky. In focus discussion groups, they voiced their individual coping experiences, for example, as follows:

> "Sea water means mother's milk to us...We cannot retaliate against anything she does. This was the first time she became angry with us. We still value her love and care. No one will hate their mother if she shows her anger once."
>
> "I would cry if I were the only one who lost my son. There were many who lost more than I did. When I tried to console others, I learnt to control my tears."
>
> "An ordinary fisherman's life has never been luxurious. We are known for our hardships and resilience. We are proud of that."
>
> "We are different from the victims of the American bomb blast (September 11, 2001). We won't expect much from our lives. If we get some food, water and shelter, we can go on with our life. That's why we did not develop any mental illness."
>
> "We were chosen by our God to survive. We have the special responsibility of rebuilding our village and to change the future."

> "We have been taught by our elders that a fisherman should always be ready to dedicate his life to the sea. We know that our life will end one day in the sea. Our elders have taught us to brave ourselves against adversities, on and off the sea."

> "Indians are tolerant to many hardships. During our freedom struggle Mahatma Gandhi, father of our nation, was hurt on several occasions. He did not lose his courage and he did not give up, so will we not..." (Rajkumar et al., 2008, pp. 847-849).

In all of the discussion groups, participants were asked to distinguish the most important coping strategy they employed. In four out of six groups, participants claimed that their religious beliefs were the most important factor enabling them to survive. Religious rituals addressing the well-being of the deceased and cultural rituals such as public mourning also facilitated coping. Moreover, participants also reported seeing positive changes in the community after the tsunami, for example, scheduled castes have a better status now due to better job opportunities. Through the formation of self-help groups, housewives were, for example, able to start a coir rope business. Now they can earn their own money and make their own decisions. Finally, most participants felt a stronger sense of unity in the community after the tsunami.

In another qualitative study examining the cultural context of coping among survivors of the 2004 tsunami in Tamil Nadu, five common ways of coping were found: (1) returning to routine, (2) rebuilding family structures, (3) communal sharing of resources, (4) emotional expression of grief and loss to a supportive listener, and (5) finding benefits from the disaster experience (Kayser, Wind, & Shanker, 2008).

In a study with Sri Lankan tsunami survivors 20 to 21 months after the tsunami, Hollifield et al. (2008) assessed the degree to which culturally relevant activities were helpful in coping with symptoms. Focus discussion groups revealed the following activities in which people engaged: (a) own strength, (b) family/friends, (c) general religious practice, (d) Ayurveda, (e) Western-model medical doctor/hospital/health system, (f) religious and cultural rituals (i.e. Thovil/Methuruma ritual healing ceremonies, (g) Bodhi-puja (a ceremony practiced by Buddhists to venerate the Bo-tree that fulfills emotional and devotional needs), and (h) horoscope (Hollifield et al., 2008, p.40). Participants were asked to report how helpful each activity was (0 = not at all to 3 = extremely). It was reported that it was either very or extremely helpful to utilize one's own strength by 70 percent of the participants, family and friends by 56 percent, a Western-style hospital by 56 percent, one's own religious practice by 53 percent, Bodhi-puja by 44 percent, Ayurveda by 21 percent, horoscope by 18 percent, and Thovil/Methuruma by 3 percent. Most coping strategies were not associated with psychological distress, only using the traditional Thovil/Methuruma was. However, the direction of causality cannot be determined using this method.

10.4 Risk and Protective Factors for Developing PTSD

When investigating the relationship between the demographic and trauma severity variables and the level of PTSD symptoms, it was found that married people were less traumatized, as well as people who have some other occupation than fishing. The people in the youngest age group were more traumatized than the other age groups. No gender effects were found and also no effects of the level of education. The variables of trauma severity were not related to the level of PTSD symptoms with the exception of being injured during the tsunami. That means individuals who have lost their family members or their houses were not more traumatized. This phenomenon seems strange at first glance. One has, however, to take into account that when people are part of a closely-knit community and form strong bonds with the other individuals within it, the loss of a neighbor or a close family member becomes equally traumatic (de Vries, 1996). The findings that trauma severity variables do not lead to higher levels of PTSD are also in contradiction to findings from other studies in India. For example, in a study with Tamil tsunami victims by Kumar et al. (2007), odds ratio analyses revealed that persons without income and women were almost three times as likely to develop PTSD. However, those individuals who were injured during the tsunami were also almost three times as likely to develop PTSD, which corresponds to this dissertation study's results. In Kumar et al.'s (2007) study, those who had lost a family member were twice as likely to develop PTSD, and those who did not receive adequate support after the tsunami nearly one and a half times.

However, somewhat different risk and protective factors were found in a Sri Lankan sample. In a recent study with tsunami victims in Sri Lanka by Levy, Slade, and Ranasinghe (2009), it was found that six months after the disaster, a pessimistic explanatory style was associated with worse health and PTSD. A belief in karma, which is typical for a mainly Buddhist population, was also associated with worse health, but not with PTSD. People with a pessimistic explanatory style attribute the cause of events as stable, global, and internal, which the authors postulated as being similar to a belief in karma. The authors conclude from this study that a belief in karma might only impact physical health, but the affect on mental health might be more varied. People might feel comforted by the belief in karma, as it presents a world of balance, or they might feel that bad fortune cannot be avoided. In another study with Buddhist participants, specifically Tibetan refugee children in Dharamsala, India, the authors report that several participants mentioned that the traumatic events they had experienced on their escape route from Tibet were the result of their karmic debts. Now they had exhausted that karma and would lead a happier life in the future. This belief system freed them from attributions of responsibility and therefore reduced feelings of guilt and anticipation of recurrence in the future (Servan-Schreiber, Le Lin, & Birmaher, 1998).

In another recent study in Sri Lanka, gender effects were also not found. The authors attribute this finding to the extremity of the event. In situations of extreme trauma, gender effects might be negligible. Moreover, age, marital status, and education were also not associated with PTSD, depression, anxiety, somatic complaints or impairment. Participants whose life was in danger, who lost a family member, or whose family member was injured, showed significantly more PTSD symptoms. Property damage, however, was not a significant predictor for PTSD (Hollifield et al., 2008).

In a literature review on public mental health in low-income countries, de Jong (2005) lists examples of protective factors that have been found to mitigate the effects of disaster and conflict in these countries as follows: (a) the presence of a social network, including a nuclear or extended family; (b) social support and self-help groups for empowerment and sharing; (c) employment or other possibilities of income generation; (d) access to human rights organizations; (e) recreation and leisure activities; (f) the possibility to perform culturally prescribed rituals and ceremonies; (g) political and religious inspiration as a source of comfort, meaning, and a perspective for the future; (h) camps with a limited size; and (i) coping skills, intelligence, and humor (p.7).

Another protective factor is the religious belief system of an individual. In a pilot study, Brune, Haasen, Krausz, Yagdiran, Bustos and Eisenman (2002) investigated to what extent strong religious beliefs had a positive effect on therapy outcome and posttraumatic adjustment. They found a strong positive connection between therapy efficiency, successful adjustment, and strong religious beliefs. Different societies encourage different ways of coping with traumatic events. Regardless of whether a society emphasizes individualism or conformity, there seems to be a universal tendency for people under threat to form very close attachments to other people. Each society has developed social and religious structures that function to help acutely distressed people until they can look after themselves again. In addition, societies provide their members with traditions, institutions, and value systems that keep the individual from becoming overwhelmed by a traumatic experience. Through religion, suffering is placed in a larger context. Religion also provides a sense of purpose and a particular solution to the uncontrollability of life experiences.

10.5 Conclusions from the Path Analysis Model

A model with latent variables was not feasible, as the subscales of the measures do not adequately represent the latent constructs. This may be the result of the scales being tailored for a Western industrialized sample. This study took place in a non-Western, non-industrialized setting, and it is open to debate whether Western psychometric scales and Western concepts of traumatic stress are applicable in such a setting and cross-culturally valid.

However, the fact that a path analysis model was feasible at all for this sample implies that the questionnaires were applicable in this non-Western, non-industrialized setting. The participants must have understood how to complete the questionnaires enough to give meaningful responses to the items. Moreover, the results that the SOC and social support significantly decrease the level of traumatic stress and being affected by the tsunami leads to more traumatic stress also confirms the validity of the scales.

10.6 Trauma from a Cultural Perspective

In every culture worldwide, people endure, recover from, and succumb to trauma. How the individual and the community perceive, appraise, and deal with a traumatic event is influenced by culture. The question arises: What is the relationship between culture and trauma?

In the DSM-IV, PTSD is the only disorder caused by an exogenous event (American Psychiatric Association, 1980, 1994). Therefore, the disorder is not ascribed to the individual alone, but also to social and cultural factors. Culture plays an important role in helping the individual deal with traumatic events by providing the context in which social and other material support can be accessed. A positive evaluation of the self in social context, as well as social support, corrects the negative effects of stressful events. In most societies, the family/kin group is chiefly responsible for its members, and the social security systems are based on filial piety and kin responsibility. After a disaster strikes, the social security or filial (kin) system responds to problematic emotional reactions and brings order to the person's emotional reactions to traumatic events. Moreover, culture provides the individual with an identity in terms of norms and values and supplies a shared vision of the future. Cultural stories, rituals, legends highlighting the mastery of communal trauma, the relationship to the spiritual realm, and religion itself help members of the community to reorganize their life. Continuity is maintained even during discontinuous and disrupting events. In this way, culture can become a buffer, so that potentially traumatizing experiences do not lead to PTSD. If the traumatized individuals lose their supportive kin network, they automatically lose their healing resources. In contrast, it is the individual's own responsibility and "duty" to be healthy in Western societies (deVries, 1996). In some circumstances, when society as a whole is affected by the trauma, the basic structure of the cultural system can be disrupted. Then homeostatic mechanisms such as rituals, social organization, and the economic system are no longer sufficient to restore a sense of safety and belonging, and need to be altered (Shalev, 1996). However, when cultural defense mechanisms against traumatization are lost, the individual is left on his or her own to gain emotional control.

People's appraisal of their level of control over the traumatic situation also varies between cultures. Viewing PTSD as a normal response to an abnormal condition implies that people usually can have control over their fate, which is a very optimistic view. In fact, only in recent history, the notion that people can shape their own destiny has emerged in Western countries. Perhaps the decline of religion has played a role in this development, bringing about people's illusion of being in charge of their own future. Eastern religions do not offer their followers such a promise. Both Hinduism and Islam teach people that life is entirely determined by fate, and that the individual has to submit his or herself to God's or Allah's will. So far, no comparative scientific research has been conducted on the prevalence of PTSD in areas of the world where these religions have a large following. Therefore, it is uncertain how such religious beliefs influence people's response to trauma.

Cultures create meaning systems that explain the causes of traumatic events. In traditional societies with concepts of an external causation of traumatic events (such as god or gods, or witchcraft), the individual's experiences of illness and trauma are directly linked to the larger society. By communicating suffering, filial responsibility and dependence are evoked. Rituals and cultural customs help individuals control their emotions, reorganize their behavior and link the sufferers more intimately to the group. For example, certain grieving rituals (e.g. highly emotional acting out of grief) move the individual from a state of shock to grief, and ultimately to non-bereavement. These rituals are designed to help bereaved people return to their normal level of functioning. If these culturally regulated processes do not take place, the reestablishment of normal functioning is less likely to occur. (deVries, 1996).

People in many developing countries perceive the role of God in disasters to be much more immediate, simple, and direct than people in industrialized nations who see Him as "benign but removed" (Bauman & Sims, 1974).

10.7 Critical Voices Against Using the PTSD Construct in a Non-Western Setting

In spite of the encouraging findings concerning the universality of the PTSD construct, some authors have voiced their critical opinions regarding the application of PTSD in non-Western settings.

As the secularization of life in Western society has taken place, traditions and religion have lost their power. Instead of explaining human suffering as the will of God, it had become a scientific problem. Human pain was considered as a technical problem that could be solved through technical intervention by experts. Thus, psychiatric knowledge was to be understood as neutral, objective, disinterested, and universally applicable (Bracken, 2002). However, according to Summerfield (2004), Western psychiatry is far from neutral or universally applicable. Rather, psychiatric science is comprised of socially constructed ideas about mental health. This includes socially constructed ideas about the outcomes, which carry a measure of self-fulfilling prophecy. Summerfield (2004) implies that labeling a community of people as at risk for PTSD, for example, after a natural disaster occurs, undermines their capacity to be resilient. An expansive mental health industry has promoted the idea that the trials of life represent noxious influences, which not only hurt the average citizen, but also disable him or her. In previous generations, Western society emphasized stoicism, composure, and self-sufficiency, but now people are invited to view a widening range of life experiences as inherently risky and liable to make them ill. This makes them less resilient and more likely to seek professional help.

Likewise, importing the concept of PTSD into non-industrialized, non-Western societies, may do more harm than good. People's responses to a disaster may be pathologized and then natural resiliency is undermined. Most survivors of traumatic events are, in fact, upset but not ill. They still function effectively in their day-to-day life and are quite resourceful.

Moreover, effective, indigenous methods of coping with disaster are marginalized if Western interventions and concepts are introduced according to Summerfield (2004). According to Higgenbotham and Marsella (1988), psychiatric universalism risks being imperialistic, reminding people of the colonial era when what was presented to the indigenous peoples was that there were different types of knowledge existed, and theirs was second-rate. Marsella et al. (1996) also warn against ethnocentricity. Virtually all of the theory, research, and measurement on PTSD have been generated by Euro-American, European, Israeli, and Australian researchers. When these concepts and approaches are applied indiscriminately to non-Western cultures, there are risks of ethnocentric bias. Ethnocentrically biased concepts do not encompass or include the experience of non-Western people, particularly with regard to their notion of health, illness, personhood, and normality, as well as their expression of symptomatology and phenomenological experiences of disorders such as PTSD (Marsella, Friedman, & Spain, 1996).

It has been argued that the stronger sense of community and spiritual connectedness seen in non-Western societies in contrast to the individualistic subjective focus found in Euro-American cultures produces crucial differences in the psychological impact and expression of traumatic encounters. The philosophy that one must accept events in life, regardless of their impact, is an important topic that requires better understanding within cultural contexts that emphasize personal control and personal responsibility. It should also be noted that trauma seems well understood by people from non-Western cultural backgrounds (Marsella, Friedman, Gerrity, & Scurfield, 1996).

Zur (1996) goes as far as to say that Western psychiatry is merely one of many ethnomedical systems and is culture-bound to Western society. The fallacy is the assumption that because phenomena can be identified in different cultural settings, they mean the same thing in those settings (Zur, 1996; Summerfield, 2004). The histories of terms such as "depression" or "PTSD," and the particular meanings and responses they mobilize, are simply not straightforwardly reproduced elsewhere. For example, in many non-Western cultures, the role of the self is completely different. The self is not a freestanding unit, but interconnected with the social group. The harmony of the family or in-group is more important than the autonomy of the individual. Social connectedness, not personal depth, is the measure of the moral value of the self (Bracken, Giller, & Summerfield, 1995, Summerfield, 2004). The Bhutanese torture victims of Shrestha et al.'s study (1998) had a completely different explanatory model for their symptoms than Western psychiatrists. They did not see any relationship between trauma and psychological problems, but rather explained psychiatric symptoms as resulting from bad karma, spirits, witchcraft, or an offended god. Therefore, it is not possible to say that one set of ideas about human pain and suffering is definitive. There is no one definitive psychology according to Summerfield (2004). Of course, many aspects of Western psychiatry or psychology, including medication, are very useful. However, it may be helpful to combine them with the indigenous healing methods of the culture at study (Summerfield, 2004).

In a qualitative study in Latin America investigating the occurrence of PTSD symptoms after violent conflicts between guerrillas and the army, Elsass (2001) also found PTSD to be a culture-related construct. In one sample with Peruvian villagers, the participants reacted with conventional PTSD symptoms, but they also reported symptoms not included in the DSM-IV criteria. Some examples are: "evil and bad thoughts", "a burning head", and "having cried the eyes out" (Elsass, 2001, p.309). Most participants acknowledged that "bad times" lead to somatic symptoms such as headaches and tiredness, but they did not believe that these symptoms should be treated with psychotherapeutic interventions. In addition, they expressed a desire to forget the past, and their traumatic memory was more collective. However, in a second sample with Colombian villagers, where people are more individualistically oriented, participants related their symptoms to the violent assaults and showed symptoms more consistent with the conventional DSM–IV criteria for PTSD. The traumatic memory was more individualized. In Peru, psychosocial work is carried out mostly by strengthening the construction of the entire community, whereas in Colombia, psychological interventions are more widely used and accepted. Interventions are carried out by psychologists and psychiatrists, as well as local healers and shamans. This shows that depending on the culture of the population under study (collectivistic or individualistic), different symptoms will be manifested and different interventions required.

According to Bracken et al. (1995), this does not mean that the concept of PTSD should be abandoned, but rather that its limitations should be recognized and its application in non-Western settings be approached with caution. In the view of these authors, recovery from traumatic events is inherently linked to the rebuilding of social and economic networks, cultural institutions, and respect for human rights. In this way, one can understand why some survivors become victims of psychological impairment, and more importantly, why the majority do not. Moreover, in de Jong's (2005) view, it is important to expand one's focus and include other disorders such as mood disorders, somatoform disorders, and other anxiety disorders in recovery programs. Psychosocial problems should also not be neglected. People in so-called "developing" nations might be suffering from economic constraints, lack of medical care and other such difficulties, which could cause more distress than the actual traumatic event, and thus bias psychological assessment aimed at finding relationships between traumatic events and PTSD (Kagee, 2005). However, that does not mean one should completely abandon the internationalization of the PTSD construct. Studies using Western concepts of mental disorders can help researchers discover how disaster relief can be improved. The results can give insights into implication for interventions

10.8 Disaster Relief within a Collectivistic Context

In a qualitative study with victims of the 2004 tsunami in Nagapattinam, Tamil Nadu, Kayser et al. (2008) investigated how the collectivistic cultural context of southern India shaped relief efforts aimed at enhancing coping of the survivors with mass level trauma. To achieve this goal, ten individuals of Indian ethnicity who provided emergency relief in the area of the Nagapattinam District were interviewed using semi-structured interviews. It was assumed that the collectivistic Indian culture would influence how people generally cope with stressors and what they would expect from relief workers.

In India, families and communities as opposed to nuclear relationships bind people together. Communities are typically structured around a common occupation such as fishing, and thus the fisherfolk live together in a community. Moreover, the women form a closely-knit community among themselves, with strong networks for mutual support. Marriage is an economic arrangement between two families, usually of the same caste. There is not much variation in the family structure, with the oldest male family member as head of the households. Aunts, uncles, cousins, and other members of the extended family show solidarity in times of need, troubles, and important occasions such as marriage, death or unemployment (Pandey, 2006). Religious values and beliefs also shape one's response to a crisis and can give meaning to an otherwise arbitrary traumatic event. In India, the majority of the population is Hindu, and thus the belief in karma is predominant. That means that what a person is experiencing now is a result of previous actions. It is important to lead a life according to dharma, or according to a sacred code of conduct. This requires the individual to perform certain rituals and to behave in a moral way to benefit oneself, one's family, and the society at large (Parish, 1994). In this way, it is one's karma or fate to have experienced the tsunami and one's dharma to take care of surviving family members.

Thus, when Kayser et al. (2008) asked the Indian emergency personnel about the most important aspects related to survivors' ability to cope after the tsunami, and which specific needs they expressed, the answers can be attributed to the specific cultural context. Villagers in the Nagapattinam District expressed their need to return to a routine or sense of normalcy, which meant women attending to routine household duties such as cooking and for the men to return to their fishing occupation. For the fishing community, the first priority was to obtain new boats and nets. Returning to their fishing helped them to cope with the remaining stressors such as living in temporary housing. Some Western organizations sent Western-type clothes, but this was not well received, as the women and girls in this region only wear the traditional sari. Moreover, the fisherfolk did not accept used clothing. Packaged food, which did not consist of the usual diet of the local people, was also not appreciated.

Moreover, it was important for the tsunami survivors to rebuild family structures. Given the high death toll of women in South India, the government encouraged widowers to remarry quickly. Staying unmarried and raising children alone would have been unthinkable. However, this created a new set of problems, as girls were married off quite young (under 18) to much older widowers, so that the family would receive financial aid.

Another important coping strategy was sharing communal resources. As is typical for a collectivistic culture, the fishing community demonstrated the cultural value of interdependence by the refusal to accept help from any organization unless everyone in the community would benefit. The villagers expressed a strong sense of responsibility to care for one another. This signals the primacy of community connectedness as a foundation for coping with a catastrophic event.

What also helped them cope with the trauma was the emotional expression of grief and loss to a supportive listener. Therefore, the relief workers reported that in the beginning, they would just sit and talk the tsunami victims.

Finally, the survivors tried to focus on the benefits that they experienced from the disaster. For example, many people received an opportunity to learn a new skill for a new job or were able to start a business through a micro-credit loan. Women in particular were addressed by the micro-credit systems, and their status is now starting to change. They currently have an opportunity to uplift their family. Another example is that people have started to improve their hygiene practices. Thus, relief efforts should not stop at fulfilling the short-term needs of the survivors, but also empower survivors for a better future. This involves interventions such as providing micro-credit loans to start businesses, as well as offering skill and job training programs. In this way, the communities will be sustained and empowered in the long run, instilling in the survivors a new sense of hope for the future (Kayser et al., 2008).

According to Rao (2006), in the rescue phase after a disaster, cultural sensitivity also needs to also be applied when dealing with dead body identification. As the Indian culture includes elaborate grieving rituals, it is important for the survivors to identify their dead family members so they can perform the appropriate funeral rites. The unceremonious disposal of corpses can hinder the recovery process of the survivors. In the relief phase, psychological first aid through trained community-level workers and volunteers needs to be provided. While survivors are still living in temporary shelters, this first aid also encompasses the normalization of routine by involving the people in routine everyday activities and the reconstruction of their houses. Informal schooling and play activities for children, as well as group prayer and mourning has been found to have a healing effect. In the rehabilitation phase, affected individuals move back into their homes. The new houses should be compatible with the people's lifestyle, and ideally they should be involved in the construction process so they will accept their new houses. In the rebuilding phase, social support networks need to be supported and new ways of earning a livelihood need to be made possible (Rao, 2006).

In an interview-based, qualitative study by Nirupama (2009) in 161 tsunami-affected villages in Tamil Nadu, Pondicherry, Andhra Pradesh, and Kerala, very few participants mentioned that the Sea Goddess was displeased and angry and thus punished them. They recognized the tsunami as a natural hazard and did not give any religious explanation. Still, the survivors all along the coast performed rituals to the Goddess Gangamma (Goddess of sea/water) within one week of the disaster in the midst of their misery and distress (Nirupama, 2009, p.5). Nirupama (2009) also discovered through these interviews that immediate help came from the nearby communities that were not affected by the disaster. This includes rescuing the injured and people trapped in unsafe places, along with the sick, children, and women. By the afternoon of December 26, 2004, official government emergency teams and NGOs provided food, safe drinking water, temporary shelters, and emergency medical assistance. The government aid workers and private teams, with the help of the locals, started clearing up the debris. The following day was dedicated to identifying the dead bodies and handing them over to the families for cremation. Then communications, electricity and sanitation needed to be provided. Each family also received monetary relief aid, as well as staple food, kerosene oil, clothing, bed sheets, and cooking utensils. However, few fishermen received new boats and nets or had their damaged ones repaired, even though their top priority was to restart their livelihood. Moreover, these fishermen actually preferred to receive their relief aid in the form of equipment rather than money, as they did not trust that the government would give them their full allotted amount. The fishing community also complained that the government relief aid overlooked the psychological aspects of the trauma. They did not arrange for professional counseling. Another shortcoming was that they failed to provide recreation facilities such as radios or television in the shelters.

However, from a report by the WHO, India (2006), it becomes clear that the government did conduct some programs for training community-level workers in psychosocial counseling. Professionals (psychiatrists, professors of social work) trained master trainers (faculty from the psychiatry department, social scientists, clinical psychologists), who in turn conducted the training of trainers (teachers, health workers, nursing and social work students, youth groups), who in turn trained community-level workers (health workers, auxiliary nurses/midwives, self-help groups, community volunteers). For example, in Kollam District, Kerala, community-level training was conducted for three days in May, 2005, in association with the Department of Psychiatry, Medical College, Thiruvananthapuram. Fifty-one community level-volunteers were trained in basic counseling skills, relaxation techniques, and ventilation and listening skills. In the Kollam District alone, the community-level workers visited 3,520 houses, screened 6,526 persons and identified 213 problem cases. The diagnoses included depression, anxiety disorder, PTSD, complicated grief, and alcohol dependence, and appropriate treatment was provided.

Nevertheless, NGOs seemed to have provided better psychological care than the government services. The response to the 2004 tsunami by two NGOs in Chennai, Tamil Nadu, is described in a paper by Vijaykumar, Thara, John, and Chellappa (2006). As psychological sequelae, these authors report that blame and guilt were seen in many tsunami victims, especially in mothers who tried to hold onto their children only to have the waves sweep them away. Women whose clothes were torn off by the waves felt extremely ashamed and had difficulty coping with this incident. Moreover, even two years after the tsunami, people were constantly afraid that the tsunami would reoccur. To ameliorate this psychological distress, the concept of community-based psychological intervention was developed. It was believed that relief workers involved in providing material relief could be trained in psychological intervention techniques, and that they would be able to provide psychological support while providing their other services. However, the voluntary relief workers who were inspired to help came from all over India and were not native to the affected areas. Thus, they could not converse in the local language, which did not matter for providing material relief, but did not enable them to conduct psychological counseling. However, the need for counseling and medication was assessed to be the fifth most important need among the survivors (the first four were household items, shelter, assistance in getting financial compensation, and clothing). Approximately 53 percent of the individuals surveyed indicated the need for some kind of psychological support. Thus, a local NGO in Chennai (Schizophrenia Research Foundation, SCARF) trained local community-level relief workers to provide crisis intervention and basic psychological counseling (Vijaykumar, et al., 2006). A typical counseling training session includes many practical tools such as information about government financial schemes and innovative suggestions about income generation. “Though counseling is very important, we must be careful about not overdoing it. When we find someone can cope, then we have no role”, states a psychiatric social worker of SCARF, according to Chatterjee, (2005). SNEHA, another NGO in Chennai working for the cause of suicide prevention, trained volunteers to conduct Training of Trainers (TOT) Program for providing psychological support for the tsunami survivors. Moreover, they set up a tent in the worst affected area in Chennai, and SNEHA's volunteers were available for three hours daily to provide emotional support and befriending. In this way, the affected people could recount their experiences and nightmares. If it was felt that more help than merely befriending was needed, two experienced professional counselors also provided counseling. In addition, sports and cultural activities were organized to bring back a semblance of normality (Vijaykumar et al., 2006).

Vijayakumar and Kumar (2008) conducted an intervention study one year after the tsunami to evaluate the mental health support provided by volunteers trained by SNEHA for the tsunami survivors who had lost at least one close family member. The volunteers essentially provided support by the process of befriending, that is, counseling based on offering human contact and emotional support. The findings indicate that in the intervention group, general psychological distress, as well as depressive and post-traumatic stress symptoms significantly decreased compared to a non-treatment control group. Regression analyses also showed that the intervention significantly predicted lower scores on the measures of psychological distress.

Another paper examined the role played by a network of 12 local NGOs (the East Coast Development Forum, ECDF) in response to the tsunami. The heart of the program was supporting already existing small self-help groups with financial aid and supplies. Self-help-groups (SHGs) are autonomous groups of 10 to 20 women who meet on a regular basis to discuss social issues, pool savings, and make decisions concerning what to do with these savings as well as bank deposits that they have made to provide micro-credits to members. It was this informal system based on trust, confidence, and credibility that enabled the ECDF to access the affected communities to determine what was required. In the first few hours after the tsunami, the local communities themselves were for the most part entirely responsible for helping each other and providing basic amenities. Next, governmental and non-governmental agencies arrived on the scene. NGOs filled the gaps not covered by government programs. Their roles were to ensure that not only the fishing communities, but also the dalits (scheduled castes) and tribal communities benefited from relief packages, which would otherwise have been neglected or actively discriminated against (Kilby, 2007). The compensation money and equipment would have only been distributed among the fishermen, but not among the lower castes who nevertheless also depended on the fishing industry for their livelihood. Moreover, widows had no right to monetary compensation, as money was only allotted to men. In the relief camps, women had no privacy and were subject to gender-based discrimination (Doss, 2005; Thangaraj, 2005). Therefore NGOS established long-term rehabilitation programs to provide means for livelihood to the women of the fishing communities; and to protect the coastal region from future disasters by e.g. reforestation (Kilby, 2007).

Becker (2006) describes the experience of professional teams from the National Institute of Mental Health and Neurosciences traveling to Nagapattinam and Cuddalore in Tamil Nadu. A three-day training program for NGO-workers, teachers, and local health care providers was conducted, who in turn trained 1050 community-level workers. Trainees were taught to identify normal and severe reactions to the disaster, which required professional intervention. In this train-the-trainer program, the following basic counseling skills were taught:

1. "Feelings must be normalized and ventilated to help survivors understand the disaster and the changes they experienced in the mind.
2. Somatic and psychological symptoms could be decreased by listening, encouraging relaxation, promoting externalization of interests, and engaging in recreational activities in a group setting.
3. Culturally appropriate proverbs, metaphors incorporated into storytelling, and cultural rituals may be used to facilitate a reinterpretation of the meaning of the disaster with attention to restoration of coping mechanisms.
4. Community cohesion would be strengthened by facilitating group activities and support groups." (Becker, 2006, p. 1397).

10.9 Disaster Aid and Rehabilitation after the 2004 Tsunami: Lessons to be Learned

In the aftermath of the 2004 tsunami, the Indian mass media promoted a strong sense of self-reliance and independence, as well as a sharp resistance to accepting aid from Western nations, especially the United States. Offers of assistance were met with resentment and skepticism by the media. Local relief aid was more acceptable than culturally insensitive and inappropriate donations by foreign countries (Letukas & Barnshaw, 2008).

> The tsunami was the first disaster that showed India has the resources to cope with immense adversity. The country turned down foreign aid [sic] – and coped such that not a single medical epidemic occurred. It was able to help itself and its neighbours. (Hindustan Times, 2005).

Well-intentioned attempts to help can easily be misunderstood as meddling, interference, or even political attempts to influence or control. Local NGOs obviously know the culture and needs of a community according to Marsella and Christopher (2004).

> "Good intentions, though essential, are not enough. To help the victims of disasters, one must understand who they are and what they need from their own perspective" (Marsella & Christopher, 2004, p. 521).

Western NGOs should exercise caution when counseling or making diagnoses, as there is a danger of pathologizing human suffering. If Western NGOs wish to help, disaster workers need to be culturally competent. The language of the victims should be used and local interpreters should be employed. According to Marsella and Chistopher (2004), the indices of mental health should be expanded to other socially deviant behaviors such as drinking, gambling, and child and spousal abuse. Moreover, cultural variations in loss, grief, and bereavement should be studied. Much could be learned that could be applied across cultures. Cultural appropriate coping might occur through ceremonies or rituals. It is also necessary to use culturally sensitive assessment methods. The most helpful intervention might be strengthening the community (Marsella & Christopher, 2004)

If one assumes that traumatic events will always lead to PTSD, and that there is a clinical need to intervene with programs and procedures developed primarily in Western cultures, one is quite prejudiced. For example, cognitive behavioral therapy is the most validated psychotherapy for PTSD in the United States, but can it be applied to victims of the 2004 tsunami who live in a non-English speaking South Asian culture? Rather, it is necessary to investigate the most efficient healing methods in that specific cultural context in order to facilitate post-traumatic recovery (Wilson, 2007).

Disaster workers may need to focus more on traditional healing methods in India for mental health. Perhaps other Western countries can also learn new intervention methods such as, for example, yoga practices. In an intervention study with survivors of the tsunami on the Andaman Islands (Telles, Naveen, & Dash, 2007), it was shown that a one-week yoga program significantly reduced self-reported anxiety, negative affect, and sleep disturbances. Moreover, the measured heart and breath rate decreased as a result of the yoga intervention program, which was conducted one month after the disaster as an initiative to introduce stress management techniques. Participants were housed in temporary shelters in the capital (Port Blair), as their houses had been destroyed. The intervention program was based on India's ancient science of yoga, which included the practice of loosening exercises, specific postures, cleansing practices, breath regulation, relaxation, and meditation. The sessions lasted for 60 minutes daily for eight days. The participants were taught a yoga program combining practices intended to act on a physical, emotional, intellectual, and spiritual level. It is derived from principles in ancient texts (Patanjali's Yoga Sutras and Taittreya Upanishad) that emphasizes that yoga should promote health on all levels (4.5).

In another indigenous healing tradition, the Ayurvedic healing tradition, restorative practices are considered to unify mind, body, and spirit within the context of social conditions (Wilson, 2007). According to Kumar, Bhurga, and Singh (2005), Ayurvedic principles imply that perfect health can only be achieved when the body, mind, and soul are in harmony with each other and with their cosmic surroundings. In this holistic view of Ayurvedic science, the social dimension is also important, and the Ayurvedic healing system describes the ways in which harmony within the society can be attained. Mental equilibrium is sought by bringing the three mental qualities of rajas (activity), sattvas (tranquility), and tamas (inertia, dullness) in harmony.

11 Conclusions

The results of this study indicate that survivors of the 2004 tsunami still suffer from traumatic stress symptoms two and a half years after the disaster. If a cut-off score of 24 is used for the IES-R, almost three fourths of the affected population still suffered from clinical PTSD symptoms at the time of data collection.

The psychometric scales used were validated for Western, industrialized samples, but the factor analyses for this sample revealed satisfactory results. Satisfactory factor solutions were found and the Cronbach's alphas were quite high. It can be assumed that in spite of cultural differences, there were no problems in using these five questionnaires in this Indian sample.

Perceived social support, in particular, decreased the level of traumatic stress, and a strong sense of coherence mitigated psychological distress. However, both social support and SOC decreased the amount of avoidance coping employed, a finding whose interpretation is open to speculation, since avoidance coping is postulated to be efficient in this collectivistic setting. In the tsunami-affected group, both coping styles are significantly positively correlated with IES-R as well as BSI scores. However, correlations do not give information about causality, and this last finding could also indicate that the more people suffer from symptoms of traumatic stress, the more they engage in coping efforts. In the total group, however, approach coping increases the level of traumatic stress and avoidance coping reduces it. Therefore, the assumption that avoidance coping is the more efficient coping strategy in non-Western, collectivistic cultures, is confirmed.

11.1 Implications for Practice

Implications for practice are that after a natural disaster, protective factors such as perceived social support should be fostered. It is, for example, important that the community support network is not disrupted and that families stay together. Moreover, Western concepts of what effective coping is cannot always be applied in non-Western settings. Therefore, indigenous coping strategies need to be explored.

Once indigenous coping strategies and resources are identified and understood, they can be fostered and encouraged as a form of prevention or intervention. Thereby, processes of salutogenesis should be emphasized, i.e. looking at resiliency and factors maintaining people's health. These health protecting factors and efficient coping styles should be strengthened (de Jong, 2005).

Strengthening protective factors and coping strategies need to be kept in mind when providing both short- and long-term disaster relief. Perceived social support needs to be increased, that means, for example, survivors of disasters need to have the feeling of being taken care of by relief workers. Creating self-help-groups is, for example, an ideal way of fostering social support among the victims. Moreover, it is crucial that families stay together and those social structures in communities are not disrupted, for example, by relocation.

11.2 Limitations of the Study

Limitations of the study are for example, that it was a cross-sectional design and therefore no assumptions about causality can be made. Moreover, the results are only valid for a specific population, that is the fishing communities in South India.

Finally, the cross-cultural validity of PTSD construct in non-Western cultures is still under discussion. The author used psychometric scales that were designed with samples of Western, individualistic cultures and it open to debate whether these scales are really applicable in non-Western cultures. There is a lack of validation data for the translated measures.

Furthermore, the sampling strategy was not probabilistic. The cluster sampling method used does not produce a real randomized sample, but one that is more similar to a convenience sample.

As with all self-report measures, bias can occur because of the way respondents complete the questionnaires. Some respondents continuously give extreme answers, and some answer continuously the middle items. Moreover, memory bias can occur when respondents are asked how they coped with the tsunami at the time of the disaster. Other problems with self-report inventories are the exaggeration or underreport of the severity of symptoms. Moreover, respondents may answer according to social desirability (Aiken, 2002).

11.3 Suggestions for Future Research

For future research, the author suggests that qualitative studies be conducted to explore indigenous coping strategies. Moreover, qualitative studies could explore indigenous ways of experiencing PTSD. Furthermore, it is suggested that future studies be conducted as longitudinal studies, as it is always difficult to assess causal relationships in a cross-sectional study. If researchers wish to generalize the results, similar studies with other indigenous, non-Western samples would be needed.

In addition to exploring more indigenous ways of coping, future research may want to use already developed coping measures based on collectivistic cultural perspectives (e.g. the Collectivistic Coping Styles by Heppner, Lee, Wong, Park, & Wong, 2006; the Cross-Cultural Coping Scale by Kuo, Roysircar, & Newby-Clark, 2006; and the Collectivistic Coping Style Measure by Moore & Constantine, 2005). Such an assessment would make it possible to examine culturally-specific coping strategies that may serve as protective factors above and beyond the coping strategies developed from an individualistic perspective.

Moreover, ethnocultural studies of PTSD offer an opportunity to identify the universal and culture-specific aspects of the PTSD experience by comparing ethnocultural group differences in the distribution, expression, and treatment of PTSD. Identifying these differences can help clinicians adjust their practices and procedures to accommodate the shared and unique aspects of the PTSD experience. The measurement of PTSD remains a serious problem because the existing instruments often do not include indigenous idioms of distress and causal conceptions of PTSD and related disorders. For example, it is widely known that many non-Western ethnic groups present symptoms somatically rather than psychologically or existentially. Since somatization symptoms are not broadly sampled in many of the PTSD measurement instruments, it is possible that important ethnocultural variations are not being considered. If researchers first define PTSD along DSM-criteria, and then search for that 'disorder,' subtypes or variations of the disorder might remain unknown (de Jong, 2005).

There is also a need for more research among different ethnocultural populations to identify the sources of strength and resiliency that somehow mediate the onset, course, and outcome of PTSD. One wonders if certain philosophical or religious beliefs, social interaction patterns, or personal dispositions and personality orientations may be critical mediators of PTSD among certain ethnocultural groups (Marsella, Friedman, & Spain, 1996).

12 References

Acosta, F.X., Nguyen, L.H., & Yamamoto, J. (1994). Using the brief symptom inventory to profile monolingual Spanish-speaking psychiatric outpatients. *Journal of Clinical Psychology, 50,* 723-726.

Aiken, L.R. (2002). *Psychological Testing and Assessment.* New York: Allyn & Bacon.

Aldwin, C.M. (2007). *Stress, coping, and development: An integrative perspective.* (2nd edition). New York: The Guilford Press.

Alexander, D. (1999). *Natural Disasters.* Dordrecht: Kluwer Academic Publishers.

Almedom, A.M. (2005). Resilience, hardiness, sense of coherence, and post-traumatic growth: All paths leading to "light at the end of the tunnel"? *Journal of Loss and Trauma, 10,* 253-265.

American Psychiatric Association. (1980). *Diagnostic and Statistical Manual of Mental Disorders, 3rd Edition.* Washington, D. C.: American Psychiatric Association.

American Psychiatric Association. (1987). *Diagnostic and Statistical Manual of Mental Disorders, 3rd Edition, Revised.* Washington, D. C.: American Psychiatric Association.

American Psychiatric Association. (1994). *Diagnostic and Statistical Manual of Mental Disorders, Fourth Edition.* Washington, D. C.: American Psychiatric Association.

Amirkhan, J.H. & Greaves, H. (2003). Sense of coherence and stress: The mechanics of a healthy disposition. *Psychology and Stress, 18* (1), 31-62.

Andreski, P., Chilcoat, H., & Breslau, N. (1998). Post-traumatic stress disorder and somatization symptoms: A prospective study. *Psychiatry Research, 79,* 131-138.

Andrews, B., Brewin, C. R., & Rose, S. (2003). Gender, social support, and PTSD in victims of violent crime. *Journal of Traumatic Stress, 16* (4), 421-427.

Andrews, L., Shelvin, M., Troop, N., & Joseph, S. (2004). Multidimensionality of intrusion and avoidance: Alternative factor models of the Impact of Event Scale. *Personality and Individual Differences, 36,* 431-446.

Antonovsky, A. (1979). *Health, stress and coping: New perspectives and physical and mental well-being.* San Francisco: Jossey-Bass.

Antonovsky, A. (1987). *Unravelling the mystery of health. How people manage stress and stay well.* San Francisco: Jossey-Bass Publishers.

Antonovsky, A. (1992). Can attitudes contribute to health? Advances. *The Journal of Mind-Body Health, 8* (4), 33-49.

Antonovsky, A. (1993). The structure and properties of the sense of coherence scale. *Social Science & Medicine, 36 (*6), 725-733.

Araya, M., Chotai, J., Komproe, I.H., & de Jong, J.T.V.M. (2007). Effect of trauma quality of life as mediated by mental distress and moderated by coping and social support among post-conflict displaced Ethiopians. *Quality of Life Research, 16*, 915-927.

Aroian, K.J. & Patsdaughter, C.A. (1989). Multiple-method, cross-cultural assessment of psychological distress. *IMAGE: Journal of Nursing Scholarship, 2,* 80-93.

Aroian, K.J., Patsdaughter, C.A., Levin, A., & Gianan, M.E. (1995). Use of the Brief Symptom Inventory to assess psychological distress in three immigrant groups. *International Journal of Social Psychiatry, 41*, 31-46.

Armenian, H.K., Morikawa, M., Melkanian, A.K., Hovanesian, A., Akiskal, K., & Akiskal, H.S. (2002). Risk factors for depression in the survivors of the 1988 earthquake in Armenia. *Journal of Urban Health: Bulletin of the New York Academy of Medicine, 79* (3), 373-382.

Armenian, H.K., Morikawa, M., Melkonian, A.K., Hovanesian, A.P., Haroutunien, N., Saigh, P.A., Akiskal, K., & Akiskal, H.S. (2000). Loss as a determinant of PTSD in a cohort of adult survivors of the 1988 earthquake in Armenia: Implications for policy. *Acta Psychiatrica Scandinavia, 102*, 58-64.

Arya, A.S. (2005). *Prevention/protection and mitigation from risk of tsunami disasters*. Government of India.

Asukai, N., Kato, H., Kawamura, N., Kim, Y., Yamamoto, K., Kishimoto, J., et al. (2002). Reliability and validity of the Japanese-Language version of the Impact of Event Scale-Revised (IES-R-J): Four studies of different traumatic events. *Journal of Nervous and Mental Disorders, 190,* 175-182.

Baguena, M., Villarroya, E., Belena, A., Amelia, D., Roldan, C., & Reig, R. (2001). Propiedades psicometricas de la version Espanola dela Escala Revisada de Impacto del Estressor (EIE-R) [Psychometric properties of the Spanish version of the Impact of Event Scale-Revised (IES-R)]. *Analisis y Modificacion de Conducta, 27,* 581-604.

Barkemeier, M. & Barkemeier, T. (2001). *Kerala mit Mumbai und Madurei.* Bielefeld: Reise Know-How Verlag

Baron, N. (2002). Community-based psychosocial and mental health services for southern Sudanese refugees in long-term exile in Uganda. In J. de Jong (Ed.), *Trauma in war and peace: Prevention, practice, and policy* (pp. 157-203). New York: Kluwer Academic/Plenum Publisher.

Baschnagel, J.S., Gudmundsdottir, B., Hawk, L.W.Jr., & Gayle Beck, J. (2009). Post-trauma symptoms following indirect exposure to the September 11th terrorist attacks: The predictive role of dispositional coping. *Journal of Anxiety Disorders 23* (7), 915-922.

Baumann, D.D. & Sims, J.H. (1974). Human response to the hurricane. In G.F. White (Ed.), *Natural hazards: Local, national and global* (pp. 25-29). New York: Oxford University Press.

Beck, A.T., Ward, C.H., Mendelsohn, M., Mock, J., & Erbaugh, J. (1961). An inventory for measuring depression. *Archives of Genereral Psychiatry*, *4*, 561-571.

Beck, J.G., Grant, D.M., Read, J.P., Clapp, J.D., Coffey, S.F., Miller, L.M., et al. (2008). The Impact of Event Scale – Revised: Psychometric properties in a sample of motor vehicle accident survivors. *Anxiety Disorders, 22*, 187-198.

Becker, S. (2006). Psychosocial care for adult and child survivors of the 2004 tsunami disaster in India. *American Journal of Public Health, 96* (8), 1397-1398.

Benight, C.C. (2004). Collective efficacy following a series of natural disasters. *Anxiety, Stress and Coping: An International Journal, 17* (4), 401-420.

Berger, M. (Ed.). (2004). *Psychische Erkrankungen. Klinik und Therapie* (2nd. ed.). München: Urban & Vogel.

Berger, M. & van Calker, D. (2004). Affektive Störungen. In M. Berger (Ed.), *Psychische Erkrankungen. Klinik und Therapie* (2nd ed.) (pp.541-636). München: Urban & Vogel.

Birbaumer, N. & Schmidt, R.F. (2006). *Biologische Psychologie (6.Auflage)*. Berlin: Springer.

Bjorck, J.P., Cuthbertson, W., Thurman, J.W., & Lee, Y.S. (2001). Ethnicity, coping, and distress among Korean Americans, Filipino Americans, and Caucasian Americans. *Journal of Social Psychology, 141* (4), 421-442.

Bolger, N., Zuckerman, A., & Kessler, R.C. (2000). Invisible support and adjustment to stress. *Journal of Personality and Social Psychology, 79* (6), 953-961.

Bortz, J. & Döring, N. (2006). *Forschungsmethoden und Evaluation für Human- und Sozialwissenschaftler (4., vollständig überarbeitete und aktualisierte Auflage).* Heidelberg: Springer.

Boss, P. (2006). *Loss, trauma, and resilience: Therapeutic work with ambiguous loss.* New York: W. W. Norton & Company.

Boulet, J. & Boss, M.W. (1991). Reliability and validity of the brief symptom inventory. *Psychological Assessment: A Journal of Consulting and Clinical Psychology, 3* (3), 433-437.

Bothmer von, M. & Fridlund, B. (2003). Self-rated health among university students in relation to sense of coherence and other personality traits. *Scandinavian Journal of Caring Science, 17,* 347-357.

Bracken, P. (2002). *Trauma: Culture, meaning and philosophy.* London: Whurr.

Bracken, P.J., Giller, J.E., & Summerfield, D. (1995). Psychological responses to war and atrocity: The limitations of current concepts. *Social Science and Medicine, 40,* 1073-1082.

Bravo, M., Rubio-Stipec, M., Canino, G.J., Woodbury, M.A., & Ribera, J.C. (1990). The psychological sequelae of disaster stress prospectively and retrospectively evaluated. *American Journal of Community Psychology, 18* (5), 661-680.

Breslau, N. (2001a). The epidemiology of posttraumatic stress disorder: What is the extent of the problem? *Journal of Clinical Psychiatry, 62* (17), 16-22.

Breslau, N. (2001b). Outcomes of post-traumatic stress disorder. *Journal of Clinical Psychiatry, 62* (17), 55-59.

Breslau, N. & Davis, G.C. (1992). Post-traumatic stress disorder in an urban population of young adults: Risk factors for chronicity. *American Journal of Psychiatry, 149,* 671-675.

Breslau, N., Davis, G.C., Andreski, P., & Peterson, E. (1991). Traumatic events and posttraumatic stress disorder in an urban population of young adults. *Archives of General Psychiatry, 48*, 216-222.

Breslau, N., Davis, G.C., Peterson, E.L. & Schultz, L. (1997). Psychiatric sequelae of post-traumatic stress disorder in women. *Archives of General Psychiatry, 54*, 81-87.

Breslau, N., Kessler, R.C., Chilchoat, H.D., Schultz, L.R., Davis, G.C., & Andreski, P. (1998). Trauma and post-traumatic stress disorder in the community. The 1996 Detroit Area Survey of Trauma. *Archives of General Psychiatry*, *55*, 626-632.

Brewin, C.R., Andrews, B. & Valentine, J.D. (2000). Meta-analysis of risk factors for post-traumatic stress disorder in trauma-exposed adults. *Journal of Consulting and Clinical Psychology*, 68 (5), 748-766.

Bronisch, T. (2005). Protokoll über Krisenintervention nach der Flutkatastrophe (Tsunami) in Phuket und Khao Lak. *Psychotherapie in Psychiatrie, psychotherapeutischer Medizin und klinischer Psychologie*, 10 (1), 5-12.

Bronisch, T., Maragkos, M., Freyer, C., Müller-Cyran, A., Butollo, W., Weimbs, R., & Platiel, P. (2006). Crisis intervention after the tsunami in Phuket and Khao Lak. *Crisis, 27* (1), 42-47.

Brune, M., Haasen, C., Krausz, M., Yagdiran, O., Bustos, E. & Eisenman, D. (2002). Belief systems as coping factors for traumatized refugees: A pilot study, *European Psychiatry, 17*, 451-458.

Brunet, A., St.-Hilaire, A., Jehel, L., & Kind, S. (2003). Validation of a French version of the Impact of Event Scale-Revised. *Canadian Journal of Psychiatry, 48,* 56-61.

Bryant, R.A., & Harvey, A.G. (1996). Visual imagery in post-traumatic stress disorder. *Journal of Traumatic Stress, 9,* 613-619.

Bryman, A. & Cramer, D. (2009). *Quantitative Data Analysis with SPSS. A Guide for Social Scientists*. East Sussex: Routledge.

Burger, L., van Staden, F. & Nieuwoudt, J. (1989). The Free State Floods: A case study. *South African Journal of Psychology, 19* (4), 205-209.

Butollo, W., Krüsmann, M., & Hagl, M. (1998). *Leben nach dem Trauma. Über den psychotherapeutischen Umgang mit dem Entsetzen*. München: J. Pfeiffer Verlag.

Butollo, W. & Hagl, M. (2003). *Trauma, Selbst und Therapie. Konzepte und Kontroversen in der Psychotraumatologie*. Bern: Verlag Hans Huber.

Byrne, B.M. (2010). *Structural Equation Modeling with AMOS: Basic Concepts, Applications, and Programming (second edition).* New York: Routledge

Canino, G., Bravo, M., Rubio-Stipec, M., & Woodbury, M. (1990). The impact of disaster on mental health: Prospective and retrospective analyses. *International Journal of Mental Health, 19* (1), 51-69.

Carballo, M., Heal, B., & Horbaty, G. (2006). Impact of the tsunami on psychological health and well-being. *International Review of Psychiatry*, 18, 217-223.

Carver, C.S. (1997). You want to measure coping but your protocol is too long: Consider the Brief COPE. *International Journal of Behavioral Medicine*, 4, 92-100.

Carver, C.S., Scheier, M.F., & Prozo, C. (1992). Conceptualizing the process of coping with health problems. In H.S. Friedmen (Ed.), *Hostility, coping, and health* (pp.167-187). Washington, D.C.:APA.

Carver, C.S., Scheier, M.F., & Weintraub, J.K. (1989). Assessing coping strategies: A theoretically based approach. *Journal of Personality and Social Psychology, 56* (2), 267-283.

Cattell, R.B. & Jaspers, J. (1967). A general plasmode for factor analytic exercises and research. *Multivariable Behavioral Research Monographs, 3,* 1-12.

Chang, E.C. (2001). A look at the coping strategies and styles of Asian Americans: Similar and different? In C.R. Snyder (Ed.), *Coping with stress: Ef-*

fective people and processes (pp.222-239). London: Oxford University Press.

Chatterjee, P. (2005). Mental health care for India's tsunami survivors. *www.thelancet.com, 365* (5), 835-836.

Chen, S.-C., Lai, Y.-H., Liao, C.T., & Lin, C.C. (2005). Psychometric testing of the Impact of Event Scale-Chinese Version (IES-C) in oral cancer patients in Taiwan. *Support Care Cancer, 13*, 485-492.

Chen, S.X. & Mak, W.W.S. (2008). Seeking professional help: Etiology beliefs about mental illness across cultures. *Journal of Counseling Psychology, 55* (4), 442-450.

Chun, C.-A., Moos, R.H., & Cronkite, R.C. (2006). Culture: A fundamental context for the stress and coping paradigm. In P.T.P. Wong & L.C.J. Wong (Eds.), *Handbook of multicultural perspectives on stress and coping* (pp.29-53). New York: Springer.

Cohen, M., Ben-Zur, H., & Rosenfeld, M.J. (2008). Sense of coherence, coping strategies, and test anxiety as predictors of test performance among college students. *International Journal of Stress Management, 15* (3), 289-303.

Collings, S.J. (1997). Child sexual abuse in a sample of South Africa women and students: Prevalence, characteristics, and long-term effects. *South African Journal of Psychology – Suid Afrikaanse Tadskrif vir Sielkunde, 27*, 37-42.

Creamer, M., Bell, R., & Failla, S. (2003). Psychometric properties of the Impact of Event Scale-Revised. *Behaviour Research and Therapy, 41,* 1489-1496.

Creamer, M., Burgess, P., & Pattison, P. (1992). Reaction to trauma: A cognitive processing model. *Journal of Abnormal Psychology, 101*, 452-459.

Cronbach, L.J. (1951). Coefficient alpha and the internal structure of tests. *Psychometrika, 16,* 297-334.

Cronbach, L.J. (1984). *Essentials of psychological testing (*4th ed.). New York: Harper Row.

Cruza-Guet, M.-C., Spokane, A.R., Brown, S.C., & Szapocznik, J. (2008). The relationship between social support and psychological distress among Hispanic elders in Miami, Florida. *Journal of Counseling Psychology, 55* (4), 427-441.

Da Costa, J.M. (1871). On Irritable Heart: A clinical study of a form of functional cardiac disorder and its consequences. *American Journal of the Medical Sciences, No. CXXI.-Jan.1871*, pp17-52.

Davidson, J.R.T., Hughes, D., Blazer, D. G., & George, L.K. (1991). Posttraumatic stress disorder in the community: An epidemiological study. *Psychological Medicine, 21,* 713-721.

Davies, M.I. & Clark, D.M. (1998). Thought suppression produces a rebound effect with analogue post-traumatic intrusions. *Behaviour Research and Therapy, 36,* 571-582.

DeGirolamo, G. & McFarlane, A.C. (1996b). The epidemiology of PTSD: A comprehensive review of the international literature. In A.J. Marsella, M.J., Friedman, E.T., Gerrity, & R.M. Scurfield (Eds.), *Ethnocultural aspects of posttraumatic stress disorder: Issues, research, and clinical applications* (pp.33-85). Washington, D.C.: American Psychological Association.

De Jong, J.T.V.M. (2002). Public mental health, traumatic stress, and human rights violations in low-income countries. In J.T.V.M. De Jong (Ed.), *Trauma, war, and violence. Public mental health in socio-cultural context* (pp. 1-91). New York: Kluwer Academic/Plenum Publishers.

De Jong, J.T.V.M. (2005). Commentary: Deconstructing critiques on the internationalization of PTSD. *Culture, Medicine and Psychiatry, 29*, 361-370.

Delahanty, D.L. & Nugent, N.R. (2006). Predicting PTSD prospectively based on prior trauma history and immediate biological responses. *Annals New York Academy of Sciences, 1071*, 27-40.

Derogatis, L.R. (1993). *Brief symptom inventory (BSI), administration, scoring, and procedures manual* (third ed.) Minneapolis: National Computer Services.

Derogatis, L.R. & Cleary, P.A. (1977). Confirmation of the dimensional structure of the SCL-90: A study in construct validation. *Journal of Clinical Psychology, 33* (4), 981-989.

Derogatis, L.R. & Coons, H.L. (1993). Self-report measures of stress. In L. Goldberger & S. Brenznitz (Eds.), *Handbook of stress. Theoretical and clinical aspects.* New York: Free Press-McMillan.

Derogatis, L.R. & Melisaratos, (1983). The Brief Symptom Inventory: An introductory report. *Psychological Medicine, 13*, 595-605.

De Vries, M.W. (1996). Trauma in cultural perspective. In B.A. van der Kolk, A.C. McFarlane & L. Weisath (Eds.), *Traumatic stress: The effects of overwhelming experience on mind, body, and society* (pp.398-413). New York: The Guilford Press.

Dilworth-Anderson, P. & Marshall, S. (1996). Social support in its cultural context. In G. R. Pierce, B.R. Sarason, & I.G. Sarason (Eds.), *Handbook of social support and the family* (pp.67-79). New York: Plenum Press.

Doss, M. (2005). The adverse impact of tsunami on women and children. In A. Muricken & U. Kumar (Eds.), *Breaking silence – Voices from the margins* (pp. 68-72). Mumbai: Vikas Adhyayam Kendra for the Voices from the Margins

Dunmore, E., Clark, D.M., & Ehlers, A. (2001). A prospective investigation of the role of cognitive factors in persistent post-traumatic stress disorder (PTSD) after physical or sexual assault. *Behaviour Research and Therapy, 39* (9), 1063-1084.

Dyregrov, A., Gupta, L., Gjestad, R., & Mukanoheli, E. (2000). Trauma exposure and psychological reactions to genocide among Rwandan children. *Journal of Traumatic Stress, 13,* 3-21.

Effect of 26th December 2004 Tsunami in Kerala Coast. Southern extent of the Alappad disaster zone in Kollam. Photos by: Joseph, M., GSI, Kerala Unit, Trivandrum (2005). Retrieved from http://www.gsi.gov.in/ph_gal/kerala/index.htm

Egle, U.T., Hoffmann, S.O., & Joraschky, P. (1996). *Sexueller Missbrauch, Misshandlung, Vernachlässigung.* Stuttgart: Schattauer.

Ehlers, A. (2000). *Posttraumatische Belastungsstörungen.* Göttingen: Hogrefe.

Ehlers, A. & Clark, D.M. (2000). A cognitive model of post-traumatic stress disorder. *Behaviour Research and Therapy, 38* (4), 319-345.

Ehlers, A., Mayou, R.A., & Bryant, B. (1998). Psychological predictors of chronic posttraumatic stress disorder after motor vehicle accidents. *Journal of Abnormal Psychology, 107* (3), 508-519.

Ehlers, A. & Steil, R. (1995). Maintenance of intrusive memories in posttraumatic stress disorder: a cognitive approach. *Behavioural and Cognitive Psychotherapy, 23,* 217-249.

Elsass, P. (2001). Individual and collective traumatic memories: A qualitative study of post-traumatic stress disorder symptoms in two Latin American localities. *Transcultural Psychiatry, 38* (3), 306-316.

Emmelkamp, J., Komproe, I.H., van Ommeren, M., & Schagen, S. (2002). The relation between coping, social support and psychological and somatic symptoms among torture survivors in Nepal. *Psychological Medicine, 32*, 1465-1470.

Eriksson, M. & Lindström, B. (2005). Validity of Antonovsky's sense of coherence scale: A systematic review. *Journal of Epidemiology and Community Health, 59,* 466-466.

Essau, C.A. & Trommsdorff, G. (1996). Coping with university-related problems: A cross-cultural comparison, *Journal of Cross-Cultural Psychology, 27* (3), 315-328.

Fairbank, J.A., Schlenger, W.E., Saigh, P.A., & Davidson, J.R.T. (1995). An epidemiological profile of post-traumatic stress disorder: Prevalence, comorbidity, and risk factors. In M.J. Friedman, D.S. Charney, & A.V. Deutsch (Eds.), *Neurobiological and clinical consequences of stress: From normal adaption to post-traumatic stress disorder*. Philadelphia/New York: Lippencott-Raven.

Flach, F. (1990). The resilience hypothesis and posttraumatic stress disorder. In M.E. Wolf (Ed.), *Posttraumatic stress disorder: Etiology, phenomenology, and treatment* (pp. 37-45). Wshington, D.C.: American Psychiatric Association.

Folkman, S. & Lazarus, R.S. (1980). An analysis of coping in a middle-aged community sample. *Journal of Health & Social Behavior, 21*, 219-239.

Folkman, S. & Lazarus, R.S. (1985). If it changes it must be a process: Study of emotion and coping during three stages of a college examination. *Journal of Personality and Social Psychology,* 54, 150-170.

Franke, G.H. (2001). Möglichkeiten und Grenzen im Einsatz der Symptom-Checkliste SCL-90. *Verhaltenstherapie und Psychosoziale Praxis*, 33, 475-485.

Freedy, J.R., Saladin, M.E., Kilpatrick, D.G., Resnick, H.S., & Saunders, B.E. (1994). Understanding acute psychological distress following natural disaster. *Journal of Traumatic Stress, 7* (2), 257-273.

Frenz, A.W., Carey, M.P., & Jorgensen R.S. (1993). Psychometric evaluation of Antonovsky's sense of coherence scale. *Psychological Assessment, 5,* 145-153

Friedman, M. J. & Marsella, A. J. (2001). Post-traumatic Stress Disorder: An overview of the concept. In A.J. Marsella (Ed.), *Ethnocultural aspects of posttraumatic stress disorder: Issues, research, and clinical applications* (pp. 11-32). Washington, D.C.: American Psychological Association.

Frommberger, U. & Maercker, A. (2006). Posttraumatische Belastungsstörung (ICD-10 F4). In U. Voderholzer & F. Hohagen (Hrsg.), *Therapie psychischer Erkrankungen. State of the Art.* München: Urban & Fischer.

Frommberger, U., Stieglitz, R.D., Straub, S., Nyberg, E., Schlickwei, W., Kuner, E., & Berger, M. (1999). The concept of "sense of coherence" and the development of post-traumatic stress disorder in traffic accident victims. *Journal of Psychosomatic Research, 46* (4), 343-348.

Fujii, S., Kato, H., & Maeda, K. (1995). A simple interview-format screening measure for disaster mental health and instrument newly developed after the great Hanshin earthquake in Japan: The screening questionnaire for disaster mental health (SQD). *Kobe Journal of Medical Science, 53*, 375-385.

Fydrich, T. & Sommer, G. (2003). Diagnostik sozialer Unterstützung. In M. Jerusalem & H. Weber (Eds.), *Psychologische Gesundheitsförderung. Diagnostik und Prävention* (pp. 79-104). Göttingen: Hogrefe.

Fydrich, T., Sommer, G., Menzel, G., & Höll, B. (1987). Fragebogen zur sozialen Unterstützung (Kurzform; SOZU-K-22). *Zeitschrift für Klinische Psychologie, 16,* 434-436.

Galea, S., Ahern, J., & Resnick, H. (2002). Psychological sequellae of the Setember 11 terrorist attacks in New York City. *The New England Journal of Medicine, 346,* 982-987.

Gerrity, E.T. & Steinglass, P. (1994). Relocation stress following natural disasters. In R.J. Ursano, B.G. McCaughtey & C.S. Fullerton (eds.) *Individual and community responses to trauma and disaster* (pp.220-247). London: Cambridge University Press.

Glass, K., Flory, K., Hankin, B.L., Kloos, B., & Turecki, G. (2009). Are coping strategies, social support, and hope associated with psychological distress among Hurricane Katrina survivors? *Journal of Social and Clinical Psychology, 28* (6), 779-795.

Goenjian, A.K., Yehuda, R., Pynoos, R.S., Steinberg, A.M., Tashjian, M., Yang, R.K., Nararian, L.M., & Fairbanks, L.A. (1996). Basal cortisol, dexamethasone suppression of cortisol, and MHPG in adolescents after the 1988 earthquake in Armenia. *American Journal of Psychiatry, 154,* 929-934.

Goldenberg, I. & Matheson, K. (2005). Inner representations, coping, and posttraumatic stress symptomatology in a community sample of trauma survivors. *Basic and Applied Social Psychology, 27* (4), 361-369.

Gordon, N. S., Farberow, N. L., & Maida, C. A. (1999). *Children & Disasters.* Philadelphia:

Brunner/Mazel.

Government of India, Ministry of Home Affairs (2005). No.32-5/2004-NDM-I. Subject: Tsunami impact (26th December 2004) in A &N Islands, UT of Pondicherry and other coastal States of Tamil Nadu, Andhra Pradesh and Kerala – Relief, Rehabilitation and other follow up Measures. *SPECIAL SITREP-27* on January 9, 2005. Retrieved from http://www.who.int/hac/crises/international/asia_tsunami/sitrep/india_govsitrep.pdf

Green, B.L. (1993). Disasters and post-traumatic stress disorder. In J.R. Davidson & E.B. Foa (Eds.), *Posttraumatic Stress Disorder: DSM-IV and Beyond* (pp.75-96). Washington, D.C.: American Psychiatric Press.

Green, B.L., Lindy, J.D., Grace, M.C., & Leonard, A.C. (1992). Chronic post-traumatic stress disorder and diagnostic comorbidity in a disaster sample. *Journal of Nervous and Mental Disease*, 180 (12), 760-766.

Green, B.L., Wilson, J.P., & Lindy, J.D. (1985). Conceptualizing post-traumatic stress disorder: A psychological framework. In C.R. Figley (Ed.), *Trauma and its wake: The study and treatment of post-traumatic stress disorder* (pp. 53-69). New York: Brunner/Maze.l

Greening, L. & Stoppelbein, L. (2007). Brief report: Pediatric cancer, parental coping style, and risk for depressive, posttraumatic stress, and anxiety symptoms. *Journal of Pediatric Psychology, 32* (10), 1272-1277.

Groß, C. & Pöhlmann, K. (2004). *Traumatisierung und Burnout im Rettungsdienst.* Lage: Jacobs.

Guðmundsdóttir, H.S., Elklit, A., & Guðmundsdóttir, D.B. (2006). PTSD and psychological distress in Icelandic parents of chronically ill children: Does social support have an effect on parental distress? *Scandinavian Journal of Psychology, 47*, 303-312.

Haan, N. (1977). Coping and Defending. New York: Academic Press.

Hamilton, M. (1959). The assessment of anxiety states by rating. *British Journal of Medical Psychology*, 32, 50-55.

Hardin, S.B., Weinrich, M., Weinrich, S., Hardin, T.L., & Garrison, C. (1994). Psychological distress of adolescents exposed to Hurricane Hugo. *Journal of Traumatic Stress, 7* (3), 427-440.

Helzer, J.E., Robins, L.N., & McEvoy, L. (1987). Post-traumatic stress disorder in the general population. *The New England Journal of Medicine, 317*, 1630-1634.

Heppner, P.P., Heppner, M.J., Lee, D.-G., Wang, Y.-W., Park, H.-J., & Wang, L-F. (2006). Development and Validation of a collectivistic coping styles inventory. *Journal of Counseling Psychology, 53,* 107-125.

Herzog, G. (2004). Psychologische Aspekte von Großschadensereignissen und Katastrophen. In A. Friedmann, P. Hofmann, B. Lueger-Schuster, M. Steinbauer & D. Vyssoki (Hrsg.), *Psychotrauma. Die Posttraumatische Belastungsstörung* (S.171-195). Wien: Springer.

Heyden, R. von der (2006). Salutogenese – Ein Konzept der Gesundheitsforschung. *Ergotherapie & Rehabilitation, 4*, 6-10.

Hidalgo, R.B. & Davidson, J.R.T. (2000). Posttraumatic stress disorder:Eepidemiology and health-related considerations. *Journal of Clinical Psychiatry (Special Issue: New strategies for the treatment of posttraumatic stress disorder), 61* (17), 5-13.

Higginbotham, N. & Marsella, A. (1988). International consultation and the homogenisation of psychiatry in Southeast Asia. *Social Science and Medicine, 27*, 553-561.

Hill,R. & Hansen, D.H. (1962). Families in Desaster. In G.W. Baker & D.W. Chapman (Eds.), *Man and Society in Desaster* (pp. 185-221). New York: Basics Books.

Hindustan Times. 2005. "An Aid to Diplomacy." *Hindustan Times*. Oct. 1, Lexis-Nexis Academic.

Hofstede, G. (1980). *Culture's consequences: Comparing values, behaviors, institutions and organizations across nations*. Beverly Hills, CA: Sage.

Höge, T. (2005). Salutogenese in der ambulanten Pflege – Zum Zusammenhang zwischen organisationalen Ressourcen, erlebter Fairness, Kohärenzsinn und der psychophysischen Gesundheit von ambulanten Pflegekräften. *Zeitschrift für Gesundheitspsychologie, 13* (1), 3-11.

Hollifield, M., Hewage, C., Gunawardena, C.N., Kodituwakku, P., Bopagoda, K., & Weerarathnege, K. (2008). Symptoms and coping in Sri Lanka 20-21 months after the 2004 tsunami. *The British Journal of Psychiatry, 192*, 39-44.

Horowitz, M.J. (1978). *Stress response syndromes*. New York: Basic Books.

Horowitz, M.J. (1986). *Stress Response Syndromes*, (2nd ed.). Northvale, NJ: Jason Aronson.

Horowitz, M.J. (1997). *Stress response syndromes* (3rd ed.). Northvale, NJ: Jason Aronson.

Horowitz, M.J., Wilner, N., & Alvarez, W. (1979). Impact of Event Scale: A measure of subjective stress. *Psychosomatic Medicine, 41* (3), 209-218.

Howard, W., Loberiza, F., Pfohl, B., Thorne, P., Magpantay, R., & Woolson, R. (1999). Initial results, reliability, and validity of a mental health survey of Mount Pinatubo disaster victims. *Journal of Nervous and Mental Disease, 187,* 661-672.

Inoue, M., Tsukano, K., Muraoka, M., Otr, F.K., & Okamura, H. (2006). Psychological impact of verbal abuse and violence by patients on nurses working in psychiatric departments. *Psychiatry and Clinical Neurosciences, 60*, 29-36.

International Federation of Red Cross and Red Crescent Societies. (2004). World Disasters Report 2004 - Chapter 8 - summary. Disaster data: Key trends and statistics. Retrieved August 8, 2007, from http://www.ifrc.org/publicat/wdr2004/chapter8.asp

International Federation of Red Cross and Red Crescent Societies. (2005). World Disasters Report 2005 – Chapter 5. Retrieved October 27, 2008, from http://www.ifrc.org/publicat/wdr2005/chapter5.asp

Iwamasa, G.Y. & Kooreman, H. (1995). Brief symptom inventory scores of Asian, Asian-American, and European-American college students. *Culture Diversification in Mental Health, 1* (2), 149-157.

Jenkins, S.R. (1997). Coping and social support among emergency dispatchers: Hurricane Andrew. *Journal of Social Behavior & Personality, 12*, 201-216.

Joannidis, G. (2006). PTBS in interkulturellem Kontext: Ist das Konzept auf Flüchtlinge und Folteropfer aus nicht-westlichen Ländern anwendbar? *Zeitschrift für Psychotraumatologie und Psychologische Medizin, 4* (1), 9-25.

John, P. & Russell, P. (2007). Validation of a measure to assess post-traumatic stress disorder: A Sinhalese version of Impact of Event Scale. *Clinical Practice and Epidemiology in Mental Health, 3* (4). Retrieved February 16, 2007, from http://www.pubmedcentral.nih.gov/articlerender.fcgi?artid=1803770

Joseph, S. (2000). Psychometric evaluation of Horowitz's Impact of Event Scale: A review. *Journal of Traumatic Stress, 13* (1), 2000.

Joseph, S., Williams, R., & Yule, W. (1995). Psychosocial perspectives on post-traumatic stress. *Clinical Psychology Review, 15* (6), 515-544.

KABANI – the Other Direction (Eds.). (2007). Fair unterwegs in Kerala/Indien – Ein Reiseleitfaden. Thiruvananthapuram: Sumesh Magalassery (Red.)

Kagee, A. (2005). Symptoms of distress and posttraumatic stress among South African former political detainees. *Ethnicity and Health, 10* (2), 169-179.

Kaiser, H.F. (1958). The varimax criterion for analytic rotation in factor analysis. *Psychometrika, 23,* 187-200.

Kaiser, H.F. (1959). Computer program for varimax rotation in factor analysis. *Educational and Psychological Measurement, 19,* 413-420.

Kaiser, H.F. (1974). An Index of Factorial Simplicity. *Psychometrika, 39,* 31-36.

Kaniasty, K.Z., Norris, F.H. & Murrell, S.A. (1990). Received and perceived social support following natural disaster. *Journal of Applied Social Psychology, 20* (2), 85-114.

Kapfhammer, H.-P., Dobmeier, P., Ehrentraut, H.-B., & Rothenhäusler, H.-B. (2001). Trauma und Dissoziation – eine neurobiologische Perspektive. *Psychotherapie in Psychiatrie, Psychotherapeutischer Medizin und Klinischer Psychologie, 6,* 114-129.

Kar, N., Mohapatra, P.K., Nayak, K.C., Pattanaik, P., Swain, S.P., & Kar, H.C. (2007). Post-traumatic stress disorder in children and adolescents one year after a super-cyclone in Orissa, India: Exploring cross-cultural validity and vulnerability factors. *BMC Psychiatry, 7:8.* Retrieved from http://www.biomedcentral.com/1471-244X/7/8.

Kayser, K., Wind, L. & Shanker, R.A. (2008). Disaster relief within a collectivist context: Supporting resilience after the tsunami in South India. *Journal of Social Service Research, 34* (3), 87-98.

Kawanishi, Y. (1995). The effects of culture on beliefs about stress and coping: Causal attribution of Anglo-American and Japanese persons. *Journal of Contemporary Psychotherapy, 25* (1), 49-60.

Kessler, R.C., Berglund, P., Demler, O., Jin, R., & Walters, E.E. (2005). Lifetime prevalence and age-of-onset distributions of DSM-IV disorders in the national comorbidity survey replication. *Archives of General Psychiatry, 62*, 593-602.

Kessler, R., Sonnega, A., Bromet, E., Hughes, M., & Nelson, C. (1995). Posttraumatic stress disorder in the National Comordity Survey. *Archives of General Psychiatry*, 52, 1048-1060.

Kilby, P. (2007). *The strength of networks: The local NGO response to the tsunami in India.* Doi:10.111/j.0361-3666.2007.01030.x, © The Author(s). Journal compilation from Overseas Development Institute, 2007. Oxford: Blackwell Publishing.

King, L.A., King, D.W., Fairbank, J.A., Keane, T.M., & Adams, G.A. (1998). Resilience/recovery factors in post-traumatic stress disorder among female and male Vietnam veterans: Hardiness, post-war social support, and additional stressful life events. *Journal of Personality and Social Psychology, 74* (2), 420-434.

Kinzie, J.D., Boehnlein, J.K., Leung, P.K., Moore, L.J., Riley, C., & Smith, D. (1990). The prevalence of post-traumatic stress disorder and its clinical significance among Southeast Asian refugees. *American Journal of Psychiatry, 147*, 913-917.

Kline, B.B. (2005). *Principles and practice of structural equation modeling* (2nd ed.). New York : Guilford Press.

Knieper, T. (2006). Die Flut im Wohnzimmer. Die Tsunami-Berichterstattung als traumatischer Stressor für die bundesdeutsche Bevölkerung. *Publizistik, 51* (1), 52-66.

Knight, B.G., Gatz, M., Heller, K., & Bengston, V.L. (2000). Age and emotional response to the Northridge Earthquake: A longitudinal study. *Psychology and Aging,* 15 (4), 627-634.

Kobasa, S.C. (1979). Stressful life events, personality, and health: An inquiry into hardiness. *Journal of Personality and social Psychology, 37* (1), 1-11.

Kobasa, S.C., Maddi, S.R., & Kahn, S. (1982). Hardiness and health: A prospective study. *Journal of Personality and Social Psychology, 42* (1), 168-177.

Kokai, M., Fujii, S., Shinfuku, N., & Edwards, G. (2004). Natural disaster and mental health in Asia. *Psychiatry and Clinical Neurosciences, 58*, 110-116.

Kolb, L.C. (1989). Chronic post-traumatic stress disorder: Implications of recent epidemiological and neurophysiological studies. *Psychological Medicine, 19,* 821-824.

Konttinen, H., Haukkala, A., & Uutela, A. (2008). Comparing sense of coherence, depressive symptoms and anxiety, and their relationships with health in a population-based study. *Social Science & Medicine, 66*, 2401-2412.Koopman, C., Classen, C., & Spiegel, D. (1996). Dissociative responses in the immediate aftermath of the Oakland/Berkley firestorm. *Journal of Traumatic Stress, 9* (3), 521-540.

Koopmann, C., Classen, C.C., & Spiegel, D. (1994). Predictors of posttraumatic stress symptoms among survivors of the Oakland/Berkley, California, fire storm. *American Journal Of Psychiatry, 151,* 888-894.

Krause, E.D., Kaltman, S., Goodman, L.A., & Dutton, M.A. (2008). Avoidant coping and PTSD symptoms related to domestic violence exposure: A longitudinal study, *Journal of Traumatic Stress, 21* (1), 83-90.

Krüsmann, M. & Müller-Cyran, A. (2005). *Trauma und frühe Intervention. Möglichkeiten und Grenzen von Krisenintervention und Notfallpsychologie.* Stuttgart: Pfeiffer bei Klett-Cotta.

Kudler, H. (2000). The limiting effects of paradigms on the concept of traumatic stress. In A. Y. Shalev, R. Yehuda & A.C. McFarlane (Eds.), *International Handbook of Human Response to Trauma* (pp. 3-10). New York: Kluwer Academic/Plenum Publishers.

Kulka, R.A., Schlenger, W.E., Fairbank, J.A., Hough, R.L., Jordan, B.K., Marmar, C.R., & Weiss, D.S. (1990). *Trauma and the Vietnam generation: Report of findings from the National Vietnam Veterans Readjustment Study.* New York: Brunner/Mazel.

Kumar, M., Bhurga, D., & Singh, J. (2005). South Asian (Indian) traditional healing: Aryuvedic, shamanic, and sahaja therapy. In R. Moodley & W. West (Eds.), *Integrating traditional healing practices in counseling and psychotherapy* (pp.505-529). Washington, D.C.: American Psychological Association.

Kumar, M.S., Murhekar, M.V., Hutin, Y., Subramanian, T., Ramachandran, V., & Gupte, M.D. (2007). Prevalence of pos-ttraumatic stress disorder in a coastal fishing village in Tamil Nadu, India, after the December 2004 tsunami. *American Journal of Public Health, 97* (1), 99-101.

Kuo, B.C.H., Roysircar, G., & Newby-Clark, I.R. (2006) Development of the cross-cultural coping scale: Collective, avoidance and engagement coping. *Measurement and Evaluation in Counseling and Development, 39,* 161-181.

Laffaye, C., Cavella, S., Drescher, K., & Rosen, C. (2008). Relationship among PTSD symptoms, social support, and support source in veterans with chronic PTSD. *Journal of Traumatic Stress, 21* (4), 394-401.

Larsson, G. (2000). Dimensional analysis of the Impact of Event Scale using structural equation modeling. *Journal of Traumatic Stress, 13*, 192-204.

Lawrence, J.W. & Fauerbach, J.A. (2003). Personality, coping, chronic stress, social support and PTSD symptoms among adult burn survivors. *Journal of Burn Care and Rehabilitation, 24,* 63-72.

Lazarus, R.S. (1999). *Stress and Emotion: A new Synthesis.* New York: Springer.

Lazarus, R.S., Averill, J.R., & Opton Jr., E.M. (1974). The psychology of coping: Issues of research and assessment. In G.V. Coelho, D.A. Hamburg & J.E. Adams (Eds.). *Coping and Adaptation* (pp.249-315). New York: Basic Books.

Lazarus, R.S. & Folkman, S. (1984). *Stress, appraisal, and coping*. New York: Springer Publishing Company.

LeDoux, J. (1998). *The emotional brain: The mysterious underpinnings of emotional life.* London: Weidenfeld & Nicolson.

Lee, P.M. & Lui, H.-T.T. (2001). Coping with intergenerational family conflicts: Comparison of Asian American, Hispanic, and European American college students. *Journal of Counseling Psychology, 48* (4), 410-419.

Lemercinier, G. (1983). *Religion and Ideology in Kerala.* Louvain-la-Neuve: Centre de Recherches Socio-Religieuses, Université Catholique de Louvain.

Letukas, L. & Barnshaw, J. (2008). A world-system approach to post-catastrophe international relief. *Social Forces, 87* (2), 1063-1087.

Levy, B.R., Slade, M.D., & Ranasinghe, P. (2009). Causal thinking after a tsunami wave: Karma beliefs, pessimistic explanatory style and health among Sri Lankan Survivors. *Journal of Religion and Health, 48*, 38-45.

Link, H.K. (2005). *Nach der großen Flut. Wie der Tsunami das Leben in „meinem" indischen Dorf veränderte*. München: Nymphenburger.

Linneweber, V. & Lantermann, E.-D. (2006). Psychologische Beiträge zur (Natur-) Katastrophenforschung. *Umweltpsychologie, 10* (1), 4-25.

Lippincott, J.A. & Mierzwa, J.A. (1995). Propensity for seeking counseling services: A comparison of Asian and American undergraduates. *Journal of American College Health, 1995,* 152-159.

Litz, B.T., Orsillo, S.M., Kaloupek, D., & Weathers, F. (2000). Emotional processing in posttraumatic stress disorder. *Journal of Abnormal Psychology, 109* (1), 26-39.

Madakasira, S. & O'Brien, K.E. (1987). Acute post-traumatic stress disorder in victims of a natural disaster. *The Journal of Nervous and Mental Disease, 175* (5), 286-290.

Maddi, S.R. (2004). Hardiness: An operationalization of existential courage. *Journal of Humanistic Psychology, 44* (3), 279-298

Maddi, S.R. & Kobasa, S.C. (1984). *The hardy executive: Health under stress.* Homewood, IL: Dow-Jones Irwin.

Maercker, A. (Ed.) (1997). Therapie *der Posttraumatischen Belastungsstörungen.* Berlin:Springer.

Maercker, A. (1998). *Posttraumatische Belastungsstörungen: Psychologie der Extrembelastungsfolgen bei Opfern politischer Gewalt.* Lengerich: Pabst.

Maercker, A. (Ed.) (2003). *Therapie der Posttraumatischen Belastungsstörungen.* 2. erweiterte und überarbeitete Auflage. Berlin: Springer.

Maercker, A. & Ehlert, U. (2001). Psychotraumatologie – eine neue Theorie- und Praxisperspektive für verschiedene medizinische Disziplinen. In: A. Maercker & U. Ehlert (Ed.), *Psychotraumatologie* (S. 11-23). Göttingen: Hogrefe-Verlag.

Maercker, A. & Schützwohl, M. (1998). Assessment of post-traumatic stress reactions: The Impact of Event Scale – Revised (IES-R) [in German]. *Diagnostica,* 44, 130-141.

Maercker, A., Schützwohl, M., & Beauducel, A. (2000). Trauma severity and initial reactions as precipitating factors for posttraumatic stress disorder and chronic dissociation. *Journal of Traumatic Stress,* 13, 651-660.

Mandelbaum, D.G. (1969). Social Organization and Planned Culture Change in India. In M.N. Srinivas (Ed.), *India's villages* (pp. 15-20). Bombay: Asia Publishing House.

Markus, H.R. & Kitayama, S. (1991). Culture and the self: Implications for cognition, emotion, and motivation. *Psychological Review, 98* (2), 224-253.

Marsella, A.J. & Christopher, M.A. (2004). Ethnocultural considerations in disasters: An overview of research, issues, and directions. *Psychiatric Clinics of North America, 27,* 521-539.

Marsella, A.J., Friedman, M.J., Gerrity, E.T., & Scurfield, R.M. (1996). Ethnocultural aspects of PTSD: Some closing thoughts. In A.J. Marsella,

M.J. Friedman, E.T. Gerrity & R.M. Scurfield (Eds.), *Ethnocultural aspects of posttraumatic stress disorder: Issues, research, and clinical applications* (pp. 529-538). Washington, D.C.: American Psychological Association.

Marsella, A.J., Friedman, M.J., & Spain, E.H. (1996). Ethnocultural aspects of PTSD: An overview of issues and research directions. In A. J. Marsella, M.J. Friedman, E.T. Gerrity & R.M. Scurfield (Eds.), *Ethnocultural aspects of posttraumatic stress disorder: Issues, research, and clinical applications* (pp. 105-129). Washington, D.C.: American Psychological Association.

Mata Amritanandamayi Math. (2007). *Amma and the Tsunami. Pray & Serve*. San Ramon:

Mata Amritanandamayi Center.

Mayou, R., Bryant, B., & Ehlers, A. (2001). Prediction of psychological outcomes one year after a motor vehicle accident. *American Journal of Psychiatry, 156* (11), 1677-1685.

McCall, G., & Resick, P. (2003). A pilot study of PTSD symptoms among Kalahari Bushmen. *Journal of Traumatic Stress, 16* (5), 445-450.

McFarlane, A.C. (1986). Long-term psychiatric morbidity after a natural disaster. Implications for disaster planners and emergency services. *The Medical Journal of Australia, 145*, 561-563.

McFarlane, A.C. (1989). The aetiology of post-traumatic morbidity: Predisposing, precipitating and perpetuating factors. *British Journal of Psychiatry, 154,* 221-228.

McFarlane; A.C. (1990). Vulnerability to post-traumatic stress disorder. In: M.E. Wolf & A.D. Mosnaim (Eds.), *Post-traumatic stress disorder: Etiology, phenomenology, and treatment* (pp. 3-20). Washington, D.C.: American Psychiatric Press.

McFarlane, A.C. (2000). On the social denial of trauma and the problem of knowing the past. In A.Y. Shalev, R. Yehuda & A.C. McFarlane (Eds.), *International handbook of human response to trauma* (pp. 11-26). New York: Kluwer Academics/Plenum Publishers.

McFarlane, A.C., Atchison, M,. & Yehuda, R. (1997). The acute stress response following motor vehicle accidents and its relations to PTSD. *Annals of the New York Academy of Sciences, 821*, 437-441.

McFarlane, A.C. & van der Kolk, B.A. (1996). Trauma and its challenge to society. In B.A. van der Kolk, A.C. McFarlane, & L. Weisath (eds.), *Traumatic stress. The effects of overwhelming experience on mind, body, and society* (pp.24-46). New York: The Guilford Press.

McMillen, C., North, C.S., Mosley, M., & Smith, E.M. (2002). Untangling the psychiatric comorbidity of posttraumatic stress disorder in a sample of flood survivors. *Comprehensive Psychiatry, 43,* 478-485,

McNally, R.J. (2004). Conceptual problems with the DSM-IV criteria for post-traumatic stress disorder. In G.M. Rosen (Ed.), *Posttraumatic stress disorder: Issues and controversies* (pp. 1-14). Chichester, West Sussex: John Wiley & Sons.

Mohanty, A. (2005). Report on tsunami. Impact of tsunami in India. Retrieved from http://www.sc99ews.com/Tsunami/Impact.htm

Moore III, J.L. & Constantine, M.G. (2005). Development and initial validation of the collectivistic coping styles measure with African, Asian, and Latin American international students. *Journal of Mental Health Counseling, 27,* 329-347.

Moos, R. (1977). *Coping with physical illness.* New York: Plenum.

Moos, R. & Schaefer, J.A. (1993). Coping, resources and processes: Current concepts and measures. In L. Goldberger & S. Breznitz (Eds.), Handbook of Stress. Toronto: The Free Press.

Murphy. S.A. (1987). Self-efficacy and social support. Mediators of stress on mental health following a natural disaster. *Western Journal of Nursing Research, 9* (1), 58-86.

Mystakidou, K., Tsilika, E., Parpa, E., Galanos, A., & Vlahos, L. (2007). Psychometric properties of the Impact of Event Scale in Greek cancer patients. *Journal of Pain and Symptom Management, 33* (4), 454-461.

Nirupama, N. (2009). Socio-economic implications based on interviews with fishermen following the Indian Ocean tsunami. *Natural Hazards, 48,* 1-9.

Nishith, P., Mechanic, M.B., & Resick, P.A. (2000). Prior interpersonal trauma: The contribution to current PTSD symptoms in female rape victims. *Journal of Abnormal Psychology, 109,* 20-25.

Norris, F.H. (1992). Epidemiology of trauma: Frequency and impact of different potentially traumatic events on different demographic groups. *Journal of Consulting and Clinical Psychology, 60,* 409-418.

Norris, F.H. (2005). *Psychosocial consequences of natural disasters in developing countries: What does past research tell us about the potential effects of the 2004 tsunami?* Retrieved July 20, 2010, from http://www.redmh.org/scripts/web2printer4.pho?lnk=0%26style=prnt/styles.css.

Norris, F.H., Friedman, M.J., Watson, P.J., Byrne, C.M., Diaz, E., & Kaniasty, K.Z. (2002). 60,000 disaster victims speak: Part I: An empirical review of the empirical literature, 1981-2001, *Psychiatry, 65* (3), 207-239.

North, C.S., Kawasaki, A., Spitznagel, E.L., & Hong, B.A. (2004). The course of PTSD, major depression, substance abuse, and somatization after a natural disaster. *The Journal of Nervous and Mental Disease, 129* (12), 823-829.

North, C.S., Smith, E.M., & Spitznagel, E.L. (1994). Post-traumatic stress disorder in survivors of a mass shooting. *American Journal of Psychiatry, 151*, 82-88.

Otto, K., Boos, A., Dalbert, C., Schöps, D., & Hoyer, J. (2006). Posttraumatic symptoms, depression, and anxiety of flood victims: The impact of the belief in a just world. *Personality and Individual Differences, 40* (5), 1075-1084.

Ouimette, P.C., Ahrens, C., Moos, R.H., & Finney, J.W. (1998). During treatment changes in substance abuse patients with posttraumatic stress disorder: The influence of specific interventions and program environments. *Journal of Substance Abuse Treatment, 15* (6), 555-564.

Oyserman, D., Coon, H.M., & Kemmelmeier, M. (2002). Rethinking individualism and collectivism: Evaluation of theoretical assumptions and meta-analyses. *Psychological Bulletin*, 128 (1), 3-72.

Ozer, E.J., Best, S.R., Lipsey, T.L., & Weiss, D.S. (2003). Predictors of posttraumatic stress disorder and symptoms in adults: A meta-analysis. *Psychological Bulletin, 129* (1), 52-7.3.

Pallant, J.F. & Lae, L. (2002). Sense of coherence, well-being, coping and personality factors: Further evaluation of the sense of coherence scale. *Personality & Individual Differences, 33* (1), 39-48.

Pandey, J. (2006). India. In J. Georges, J.W. Berry, F.J.R. van de Vijver, C. Kağitçibaşi, & Y. H. Poortinga (Eds.), *Families across cultures. A 30-nation psychological study (pp. 362-369).* Cambridge: Cambridge University Press.

Parish, S.M. (1994). *Moral knowing in a Hindu sacred city: Exploration of mind, emotion and self.* New York: Columbia University Press.

Pat-Horenczyk, R., Qasrawi, R., Lesack, R., Haj-Yahia, M., Peled, O., Shaheen et al. (2009). Posttraumatic symptoms, functional impairment, and coping among adolescents on both sides of the Israeli-Palestinian conflict: A cross-cultural approach. *Applied Psychology: An International Review, 58* (4), 688-708.

Pearlin, L.I. & Schooler, C. (1978). The structure of coping. *Journal of Health and Social Behavior, 19*, 2-21.

Phifer, J.F. & Norris, F.H. (1989). Psychological symptoms in older adults following natural disaster : Nature, timing, duration, and course. *Journal of Gerontology : Social Sciences, 44* (6), 207-217.

Pietrzak, R.H., Johnson, D.C., Goldstein, M.B., Malley, J.C., & Southwick, S.M. (2009). Psychological resilience and post-deployment social support protect against traumatic stress and depressive symptoms in soldiers returning from operations enduring freedom. *Depression and Anxiety, 26*, 745-751.

Pitman, R.K., Shin, L.M., & Rauch, S.L. (2001). Investigating the pathogenesis of post-traumatic stress disorder with neuroimaging. *Journal of Clinical Psychiatry, 62* (suppl 17), 47-54.

Punamäki, R.-L., Salo, J., Komproe, I., Qouta, S., El-Masri, M. ,& De Jong, J.T.V.M. (2008). Dispositional and situational coping and mental health among Palestinian political ex-prisoners. *Anxiety, Stress, & Coping, 21* (4), 337-358.

Rajkumar, A.P., Premkumar, T.S., & Tharyan, P. (2008). Coping with the Asian tsunami: Perspectives from Tamil Nadu, India on the determinants of resilience in the face of adversity. *Social Science & Medicine, 67*, 844-853.

Rao, K. (2006). Lessons learnt in mental health and psychosocial care in India after disasters. *International Review of Psychiatry, 18* (6), 547-552.

Rao, K., Prabhu, G.G., & Subhakrishnan, D.K. (1989). Development of Coping Checklist. *Indian Journal of Psychiatry 32* (2), 128-133.

Raymond, N. (2000). The trauma of war in Sierra Leone. *The Lancet, 355*, 2067-2068.

Resick, P.A. (2003). *Stress und Trauma: Grundlagen der Psychotraumatologie*. Bern: Verlag Hans Huber.

Revel, J.P. (1996). Natural disasters and other accidents. Provisions of psychological support. The contribution of non-governmental organizations. In Y. Danieli, N.S. Rodley & L. Weisaeth (Eds.), *International responses to traumatic stress: Humanitarian, human rights, justice, peace, and development contributions, collaborative actions, and future initiatives* (pp. 289-305). Amityville, New York: Baywood Publishing Company, Inc.

Reyes, G. (2006). Psychological first aid: Principles of community-based psychosocial support. In G. Reyes & G.A. Jacobs (Eds.), *Handbook of international disaster psychology. Vol. 2.:Practices and programs* (pp1-12). Westport, CT: Praeger Publishers.

Robins, L.N., Fischbach, R.L., Smith, E.M., Cottler, L.B., Solomon, S.D., & Goldring, E. (1986). Impact of disaster on previously assessed mental health. In J. Shore (Ed.), *Disaster Stress Studies: New Methods and Findings* (pp. 22-48). Washington, D.C.: APA Press.

Roth, G. & Münte, T.F. (2003). Neurobiologische Grundlagen psychischer Traumatisierung. In G.H. Seidler, P. Laszig, R.R. Micka & B.V. Nolting

(Eds.), *Aktuelle Entwicklungen in der Psychotraumatologie, Theorie – Krankheitsbilder – Therapie* (S. 9-34). Gießen: Psychosozial.

Ruchkin, V., Schwab-Stone, M., Jones, S., Cicchetti, D.V., Koposov, R., & Vermieren, R. (2005). Is post-traumatic stress in youth a culture-bound phenomenon? A comparison of symptom trends in selected U.S. and Russian communities. *American Journal of Psychiatry,* 162 (3), 538-544.

Russell, S., Subramania, B., & Russel, P.S.S. (2004). Psychometric properties of Tamil version of Impact of Event Scale for adolescents. *International Journal of Disaster Medicine, 2,* 148-151.

Russoniello, C.V., Skalko, T.K., O'Brien, K., McGhee, S.A., Bingham-Alexander, D., & Beatley, J. (2002). Childhood pos-ttraumatic Stress disorder and efforts to cope after Hurricane Floyd. *Behavioral Medicine, 28,* 61-71.

Sack, W.H., Seeley, J.R., & Clarke, G.N. (1997). Does PTSD transcend cultural barriers? A study from the Khmer Adolescent Refugee Project. *Journal of the American Academy of Child & Adolescent Psychiatry, 36,* 49-54.

Sandell, R., Blomberg, J., & Lazar, A. (1998). The factor structure of Antonovsky's sense of coherence scale in Swedish clinical and non-clinical samples. *Personality and Individual Differences, 24,* 701-11.

Sapolsky, R.M., Uno, H., Robert, C.S., & Finch, C.E. (1990). Hypocampal damage associated with prolonged glycocorticoid exposure in primates. *Journal of Neuroscience, 10,* 2897-2902.

Scaer, R.C. (2001). The neurophysiology of dissociation and chronic disease. *Applied Psychophysiology and Biofeedback, 26* (1), 73-91.

Schnibben, C. (Hrsg.) (2005). *Tsunami. Geschichte eines Weltbebens.* München: Deutsche Verlags-Anstalt.

Schumm, J.A., Briggs-Phillips, M., & Hobfoll, S.E. (2006). Cumulative interpersonal traumas and social support as risk and resilience factors in predicting PTSD and depression among inner-city women. *Journal of Traumatic Stress, 19* (6), 825-836.

Schüssler, P. (2003). Schutzfaktoren vor PTB. Welche Rolle spielt der Kohärenzsinn? In E. Martens & K. Gusek (Eds.), *„Ans rettende Ufer". Referateband des 5. Bundeskongresses für Notfallseelsorge und Krisenintervention* (p. 177-186). Frankfurt a. M.: Verlag für Polizeiwissenschaft.

Schützwohl, M., Maercker, A., & Manz, R. (1999). Long-term post-traumatic stress reactions, coping, and social support. A structural equation model in a group of former political prisoners. In A. Maercker, M. Schützwohl & Z. Solomon (Eds.), *Post-traumatic stress disorder. A lifespan developmental perspective.* Seattle: Hogrefe & Huber.

Schwarzer, R. & Schwarzer, C. (1996). A critical survey of coping instruments. In M. Zeidner & N.S. Endler (Eds.), *Handbook of coping: Theory, research and application* (pp. 107-132). New York: Wiley.

Sek, H. & Tasikowski, T. (2002). Stressbewältigung im Rahmen der Salutogenese. In J. Schumacher, K. Reschke & H. Schröder (Eds.), *Mensch unter Belastung. Erkenntnisfortschritte und Anwendungsperspektiven der Stressforschung* (pp. 20-43). Frankfurt a. Main: VAS Verlag für Akademische Schriften.

Servan-Schreiber, D., Le Lin, B. R.N., & Birmaher, B. (1998). Prevalence of posttraumatic stress disorder and major depressive disorder in Tibetan refugee children. *Journal of the American Academy of Child & Adolescent Psychiatry, 37* (8), 874-879.

Seyle, H. (1988). Creativity in basic research. In F. Flach (Ed.), *The creative mind* (pp.255-259). Buffalo, NY: Bearly Press.

Shalev, A.Y, (1996). Stress versus traumatic stress. From acute homeostatic reactions to chronic psychopathology. In B.A. van der Kolk, A.C. McFarlane & L. Weisath (Eds.), *Traumatic stress. The effects of overwhelming experience on mind, body, and society* (pp. 77-101). New York: The Guilford Press.

Shalev, A.Y. (2000).Posttraumatic stress disorder: Diagnosis, history and life-course. In D.J. Nutt, J.R. Davidson & J. Zohar (Eds.), *Posttraumatic stress disorder: Diagnosis, management and treatment*. London: Martin Dunitz.

Shalev, A.Y. (2001). What is Posttraumatic Stress Disorder? *Journal of Clinical Psychiatry,* 62 (suppl 17), 4-10.

Shresta, N.M., Sharma, B., Van Ommeren, M., Regmi, S., Makaju, R., Komproe et al. (1998). Impact of torture on refugees displaced within the developing world. Symptomology among Bhutanese refugees in Nepal. *Journal of the American Medical Association, 280,* 443-449.

Singh, J.P. (2005). The contemporary Indian family. In B.N. Adams & J. Trost (Eds.), *Handbook of world families* (pp. 129-166). Thousand Oaks: Sage.

Smith, E.M. & North, C.S. (1988). Aftermath of a disaster: Psychological response to the Indianapolis Ramada jet crash. *Quick Response Research Report # 23.* Boulder, CO: Natural Hazards Research and Application Information Center.

Smith, E.M., Robins, L.N., Przybeck, T.R., Goldring, E., & Solomon, S.D. (1986). Psychological consequences of disaster. In J. Shore (Ed.), *Disaster Stress Studies: New Methods and Findings* (pp. 50-76). Washington, D.C.: APA Press.

Smith, P., Perrin, S., Dyregov, A., & Yule, W. (2003). Principal component axis of the impact of event scale with children at war. *Personality and Individual Differences, 34*, 315-322.

Smith, S.M. (1983). Disaster: Family disruption in the wake of natural disaster. In: C.R. Figley & H.I. McCubbin (Eds.), *Stress and the family. Vol. II: Coping with catastrophe* (pp120-147). New York: Brunner/Mazel.

Solomon, Z., Mikulincer, M. & Avitzur, E. (1988). Coping, locus of control, social support, and combat-related post-traumatic stress disorder – a prospective study. *Journal of Personality and Social Psychology, 55,* 279-285.

Somer, E., Maguen, S., Or-Chen, K., & Litz, B.T. (2007). Managing terror: Differences between Jews and Arabs in Israel. *International Journal of Psychology, 1 9, iFirst Article*

Sommer, G. & Fydrich, T. (1989). *F-SOZU. Fragebogen zur Sozialen Unterstützung.* Soziale Unterstützung, Diagnostik, Konzepte, F-SOZU (DGVT Materiale Nr. 22). Tübingen: Deutsche Gesellschaft für Verhaltenstherapie.

Sommer, G. & Fydrich, T. (1991). Entwicklung und Überprüfung eines Fragebogens zur sozialen Unterstützung (F-SOZU). *Diagnostica, 37* (2), 160-178.

Söndergaard, H.P., Ekblad, S., & Theorell, T. (2003). Screening for post-traumatic stress disorder among refugees in Stockholm. *Nord J Psychiatry, 57*, 185-189.

Sprang, G. & LaJoic A.S. (2009). Exposure, avoidance, and PTSD among Hurricane Katrina evacuees. *Traumatology, 15* (2), 10-19.

Spurrell, M.T. & McFarlane, A.C. (1993). Post-traumatic stress disorder and coping after a natural disaster. *Social Psychiatry and Psychiatric Epidemiology, 28,* 194-200.

Srinivas, M.N. (1969a). Introduction. In M.N. Srinivas (Ed.), *India's villages* (pp. 1-14). Bombay: Asia Publishing House.

Srinivas, M.N. (1969b). The Social Structure of a Mysore Village. In M.N. Srinivas (Ed.), India's Villages (pp. 21-35). Bombay: Asia Publishing House.

Steil, R. & Ehlers, A. (1998). Die Posttraumatische Belastungsstörung. In H. Reinecker (Ed.), *Lehrbuch der Klinischen Psychologie* (3. und erweiterte Auflage). Göttingen: Hogrefe.

Steinglass, P. & Gerrity, E. (1990). Natural disasters and post-traumatic stress disorder: Short-term versus long-term recovery in two disaster-affected communities. *Journal of Applied Social Psychology, 20,* 1746-1765.

Summerfield, D. (2004). Cross-cultural perspectives on the medicalization of human suffering. In G.M. Rosen (Ed.), *Post-traumatic stress disorder: Issues and controversies* (pp. 233-245). Chichester, West Sussex: John Wiley & Sons.

Sundin, E.C., Mardi, J., & Horowitz, M.J. (2003). Horowitz's Impact of Event Scale evaluation of 20 years of use. *Psychosomatic Medicine, 65,* 870-876.

Sunidharan, C.S. (2006). *Mental health risk factors and coping skills: A psychological study on survivors of tsunami victims.* Unpublished manuscript, Amrita Institute of Medical Sciences, Healthcare, Education & Research, Department of Neurology, Cochin, Kerala

Suominen, S., Blomberg, H., Helenius, H., & Koskenvuo, M. (1999). Sense of coherence and health – does the association depend on resistance resources? A study of 3115 adults in Finland. *Psychology and Health, 14,* 937-948.

Tang, S.T. & Li, C.-Y. (2008). The important role of sense of coherence in relation to depressive symptoms for Taiwanese family caregivers of cancer patients at the end of life. *Journal of Psychosomatic Research, 64,* 195-203.

Taylor, S.E., Sherman, D.K., Kim, H.S., Jarcho, J., Takagi, K. & Dunagan, M.S. (2004). Culture and social support: Who seeks it and why? *Journal of Personality and Social Psychology, 87* (3), 354-362.

Telles, S., Naveen, K.V., & Dash, M. (2007). Yoga reduces symptoms of distress in tsunami survivors in the Andaman Islands. *Advance Access Publication eCAM, 4* (4), 503-509.

Telles, S., Singh, N., & Joshi, M. (2009). Risk of post-traumatic stress disorder and depression in survivors of the floods in Bihar, India. *Indian Journal of Medical Science, 63,* 330-334.

Terheggen, M.A., Stroebe, M.S,. & Kleber, R.J. (2001). Western conceptualizations and Eastern experience: A cross-cultural study of traumatic stress reactions among Tibetan refugees in India. *Journal of Traumatic Stress, 14* (2), 391-403.

Thangaraj, M. (2005). Tsunami: Women and children. In A. Muricken & U. Kumar (Eds.), *Breaking Silence – Voices from the margins* (pp. 73-85). Mumbai: Vikas Adhyayam Kendra for the Voices from the Margins

The Official Web Portal of the Government of Kerala. (2010, September). *Cultural Background.* Retrieved September 2010 from http://www.Kerala.gov.in

The World Bank (2005). India: Tsunami Reconstruction to Cost $ 1.2 Billion, Says World Bank-ADB-UN Joint Assessment Report (on-line report on

March 14, 2005, Press Release No: 2005/380/SAR). Retrieved from http://web.worldbank.org/WBSITE/EXTERNAL/NEWS/0,,contentMDK:20392153~menuPK:34463~pagePK:64003015~piPK:64003012~theSitePK:4607,00.html

Tiet, Q.Q., Rosen, C., Cavella, S., Moos, R.H., & Finney, J.W. (2006). Coping, symptoms, and functioning outcomes of patients with posttraumatic stress disorder, *Journal of Traumatic Stress, 19* (6), 799-811.

Togari, T., Yamazaki, Y., Takayama, T.S., Yamaki, C.K., & Nakayama, K. (2008). Follow-up study on the effects of sense of coherence on well-being after two years in Japanese university undergraduate students. *Personality and Individual Differences, 44,* 1335-1347.

Triandis, H.C. (1995*). Individualism & collectivism.* Boulder, CO: Westview Press.

Tsunami Response Watch. (2005). "Tsunami in India. 6 Months After". Retrieved August 8, 2007 from http://www.achr.net/000ACHRTsunami/India%20TS/Tsunami%20India.htm

U.S. Geological Survey (USGS). Science for a Changing World. (2008). Earthquake Hazards Program. Magnitude 9.1 – Off the West coast of Northern Sumatra, Retrieved on 16 July, 2008, from http://earthquake.usgs.gov/eqcenter/eqinthenews/2004/usslav/#summary

Vaillant, G. (1977). *Adaptation to Life.* Boston: Little, Brown.

van der Kolk, B.A. (1996). The black hole of trauma. In B.A. van der Kolk, A.C. McFarlane & L. Weisath (Eds.), *Traumatic stress. The effects of overwhelming experience on mind, body, and society* (pp.3-23). New York: The Guilford Press.

van der Kolk, B.A. (2000). Trauma und Gedächtnis. In B.A. van der Kolk, A.C. McFarlane & L. Weisaeth (Hrsg.), *Traumatic Stress. Grundlagen und Behandlungsansätze. Theorie, Praxis und Forschungen zu posttraumatischem Streß sowie Traumatherapie* (S. 221-240). Paderborn, Junfermann Verlag

Vaux, A. (1988). *Social support: Theory, research, and intervention*, New York: Praeger.

Vernberg, E.M., La Greca, A.M., Silverman, W.K., & Prinstein, M.J. (1996). Prediction of post-traumatic stress symptoms in children after Hurricane Andrew. *Journal of Abnormal Psychology, 105* (2), 237-248.

Vijayakumar, L. & Kumar, S. (2008). Trained volunteer-delivered mental health support to those bereaved by Asian tsunami. An Evaluation. *Intenational Journal of Social Psychiatry, 54*, 293-302.

Vijaykumar, L., Thara, R., John, S., & Chellappa, S. (2006). Psychosocial interventions after tsunami in Tamil Nadu, India. *International Review of Psychiatry, 18* (3), 225-231.

Wang, X., Gao, L., Shinkfuku, N., Zhang, H., Zhao, C., & Shen, Y. (2000). Longitudinal study of earthquake-related PTSD in a randomly selected community sample in North China. *American Journal of Psychiatry, 157,* 1260-1266.

Weiss, D.S. (2007). The Impact of Event Scale: Revised. In J.P. Wilson & C. S.-K. Tang (Eds.), *Cross-cultural Assessment of Psychological Trauma and PTSD (*pp. 219-238*).* New York: Springer.

Weiss, D.S. & Marmar, C.R. (1996). The impact of event scale-revised. In J.P. Wilson & T.M. Keane (Eds.), *Assessing psychological trauma and PTSD* (pp. 399-411). New York: Guilford.

Weltgesundheitsorganisation (1992*). Internationale Klassifikation psychischer Störungen: ICD-10 Kapitel V(F). Klinisch-diagnostische Leitlinien.* Deutsche Übersetzung und Herausgabe von Dilling, H., Mombour, W. & Schmidt, M.H.. Bern, Göttingen, Toronto, Seattle: Verlag Hans Huber.

Wethington, E. & Kessler, R.C. (1986). Perceived support, received support, and adjustment to stressful life events. *Journal of Health and Social Behavior,* 27 (March), 78-89.

Wilson, J.P. (2007). The lense of culture. In J.P. Wilson & C.S.-K. Tang (Eds.), *Cross-cultural assessment of psychological trauma and PTSD* (pp. 3-30). New York: Springer

Witruk, E., Reschke, K. & Stueck, M. (2009). Psychische Belastung von Notfallhelfern. Auswirkungen und psychologische Notfallhilfen für Opfer der Tsunami-Flutkatastrophe in Sri Lanka und Indonesien. *Trauma und Gewalt (3. Jahrgang), Heft 3/*2009, S. 196-208.

World Health Organization. (1992). *ICD-10. The ICD-10 Classification of Mental and Behavioural Disorders. Clinical descriptions and diagnostic guidelines*. Geneva: World Health Organization.

World Health Organization. (2005). "Health action in crisis. Tsunami: India situation update" Retrieved from http://www.who.int/hac/crises/international/asia_tsunami/ind/2005_01_09/en/

World Health Organization, India. (2006). Report of the psychosocial support programme in tsunami-affected populations in India. *International Review of Psychiatry, 18* (3), 299-308.

Wu, K.K. & Chan, K.S. (2003). The development of the Chinese version of the Impact of Event Scale-Revised (CIES-R). *Social Psychiatry and Psychiatric Epidemiology, 3,* 94-98.

Yaswi, A. & Haque, A. (2008). Prevalence of PTSD symptoms and depression and level of coping among the victims of the Kashmir Conflict. *Journal of Loss and Trauma, 13*, 471-480.

Yehuda, R. (2000). Neuroendocrinology. In D.J. Nutt, J.R. Davidson & J. Zohar (Eds.), *Posttraumatic stress disorder: Diagnosis, management and treatment.* London: Martin Dunitz.

Yehuda, R. (2001). Biology of posttraumatic stress disorder. *Journal of Clinical Psychiatry, 62* (suppl 17), 41-46.

Yehuda, R. (2002). Posttraumatic stress disorder. New *England Journal of Medicine, 346,* 108-114.

Yeomans, P.D. & Forman, E.M. (2009). Cultural factors in traumatic stress. In S. Eshun & R.A.R. Gurung (Eds.), *Culture and mental health. Sociocultural influences, theory, and practice.* Malden, MA: Wiley-Blackwell.

Yeomans, P.D., Herbert, J.D., & Forman, F.M. (2008). Symptom comparision across multiple solicitation methods among Burundians with traumatic event histories. *Journal of Traumatic Stress, 21*, 305-317.

Yoshihama, M. (2002). Battered women's coping strategies and psychological distress: Differences by immigration status. *American Journal of Community Psychology, 30* (3), 429-452.

Young, A. (2000). An alternative history of traumatic stress. In A.Y. Shalev, R. Yehuda & A.C. McFarlane (Eds.), *International handbook of human response to trauma* (pp. 51-66). New York: Kluwer Academics/Plenum Publishers.

Yule, W. (2001). Post-traumatic stress disorder in the general population and in children. *Journal of Clinical Psychiatry, 62* (17), 23-28.

Zilberg, N. J., Weiss, D.S., & Horowitz, M. J. (1982). Impact of Event Scale: A cross-validation study and some empirical evidence supporting a conceptual model of stress response syndromes. *Journal of Consulting and Clinical Psychology, 50* (3), 407-414.

Zimmermann, E. (2005). Begleitung in der Not. Ein Monat als Helferin in Thailand und Indonesien. *Psychoscope*, 26 (3), 6-8.

Zimprich, D., Allemand, M., & Hornung, R. (2006). Measurement invariance of the abridged Sense of Coherence Scale in adolescents. *European Journal of Psychological Assessment*, 22 (4), 280-287.

Zur, J. (1996). From PTSD to voices in context: From an "experience-far" to an "experience-near" understanding of responses to war and atrocity across cultures. *International Journal of Social Psychiatry, 42 (*4), 305-317.

13 Appendix: Example of Efficient Disaster Relief Aid for Long- and Short-Term Rehabilitation

To demonstrate an example of efficient immediate and long-term disaster relief aid, the response of the Mata Amritanandamayi Math (MAM) to the tsunami in Alappad District, Kerala, is described. Immediately after the disaster struck Kerala's coastline, the MAM evacuated an entire village to the mainland. The villagers were taken across the backwaters on skiffs or boats. As Alappad is a peninsula running parallel with the coast (the Arabian Sea to the west and the Kayamkulam backwaters to the east), all of Alappad remained submerged for several hours. All of the evacuees were accommodated at the MAM's Amrita University, where the volunteers started cooking for 10,000 people. The MAM's educational institutions accommodated 4,000 families. The MAM supplied all of its relief centers, as well as those of others, with round-the clock services, including food, water, clothes, sleeping mats, blankets, soap, toothbrushes and toothpaste. All meals were provided to the families housed in the Amrita University campuses, as well as the many evacuees from other camps that had run out of food, including the police, the army and other disaster relief personnel. Seven fully equipped ambulances with medical staff from the Amrita Institute of Medical Sciences (AIMS), the MAM's state-of-the-art charitable hospital in Cochin, made regular rounds of MAM and non-MAM relief centers in Kollam District, Kerala. Five more ambulances were sent to affected areas in Tamil Nadu (Mata Amritanandamayi Math, 2007).

The day after the tragedy, the people staying in the relief camps were in a state of utter shock and desolation. Some were silent, almost catatonic; others wailed in emotional pain. Many did not even know that they had lost their children or relatives. Some had died from being knocked unconscious and then drowning in the water, whereas others died when a tree or piece of a house fell on them (Mata Amritanandamayi Math, 2007).

When the government allowed the villagers to return to Alappad, there was still no electricity or running water. In order to provide food and water for everyone, the MAM set up a couple dozen food counters along the main road of the peninsula, providing food and water three times a day. This service was continued for four months. Otherwise there would have been riots among the people. In the end, the MAM served a total of 7.5 million meals as part of its tsunami-relief program.

Three days after the tsunami struck, the MAM erected temporary shelters for 550 families in Kerala and Tamil Nadu. All the shelters were equipped with electric lights, running water and a television.

Long-term medical aid was also provided in the weeks and months after the disaster. The MAM's AIMS hospital in Cochin treated 900 tsunami victims, of which 450 underwent major surgeries, including 15 cardiac operations. Four hundred eyeglasses were distributed to those who lost theirs during the tsunami. Some village women who had lost their children in the tsunami had undergone fallopian-tube ligations as permanent contraception. They were very depressed and contemplated suicide. Therefore, the MAM sponsored the reversal of their contraceptive surgeries, and four of the six women elected to have the surgery delivered children.

Furthermore, a team of doctors, psychologists and social workers from AIMS provided free assistance to tsunami victims suffering from PTSD, as well as other physical, psychological, social and financial problems. Volunteers played and sang songs with the children and psychologists provided counseling. The doctors requested that the children draw pictures of their experiences so they could express themselves and discuss their tragic experiences. Many of them drew pictures of flooded villages and people clinging to coconut trees in order to save themselves. Apart from psychological counseling, the children also received educational kits (pens, books) and free tutoring so that their school education would not suffer. In April and May of 2005, two five-day camps were held and attended by approximately 10,000 children from the Alappuzha and Kollam Districts of Kerala. These camps included classes in yoga, spoken English, Sanskrit and personality development. The aim of the camps was to stimulate the children's minds in positive ways in order to help them overcome their traumatic experiences.

As many villagers' savings in the form of gold jewelry and cash were washed away in the tsunami, they were not able to pay for their daughters' weddings. Therefore, the MAM financed and conducted marriage ceremonies for hundreds of tsunami-affected couples in Kerala, Tamil Nadu and Pondicherry, providing all of the essential items such as gold necklaces and brocade saris.

The MAM also immediately began the task of building new boats for the out-of-work fishermen. The goal was to enable them to earn their livelihood again as soon as possible, which was of great emotional importance to them. One boat can support approximately seven families.

The women were encouraged to form self-help groups. The goal of these groups was for the villages to become self-reliant so they could produce all of the main items they needed by themselves. Consequently, the village women created cottage industries, including organic-banana farming, handicrafts fashioned from coconut fronds and shoe-making. The goal was for every family to have at least one member who earns 100 rupees every day to cover the family's basic needs.

In addition, the MAM trained 1,500 women in the art of tailoring and provided them with sewing machines. Free vocational training was offered to 2,500 young people in Kerala and Tamil Nadu. They were trained in carpentry, electrical work, masonry, mechanics, nursing, paramedical assistance, security, and as teachers or drivers.

In this way, the MAM did everything possible to take care of the tsunami victims' immediate and long-term needs. In fact, not only were their basic needs met, but the villages' economic situation also improved as a result of the tsunami and the relief aid that followed. The new houses are much better than the old ones, and the inhabitants of the coastal villages have better job opportunities due to the vocational training provided by the MAM.

In order to protect the coastline of Kerala, a tree-planting project was launched. By working with village leaders, the MAM has planted 90,000 casuarina saplings in Alappad. The trees serve as breakers, reducing the impact of ocean waves against the shoreline while holding the sand in place.

By September 2007, the MAM had completed 4,500 of its pledged 6,200 tsunami-relief houses. In every location, the houses were built according to government standards and have been given away free of charge to the affected people. Not only houses were built in Kerala, Tamil Nadu, Pondicherry, Andaman and Nicobar Islands, but also roads, wells, schools, community halls and healthcare centers.

At the time of the tsunami, there was only one bridge along the entire peninsula, located on Alappad's southern tip. It took three hours to evacuate the peninsula, so the MAM decided to build an escape bridge across the backwaters. In December 2006, the footbridge was inaugurated. Now 15,000 people can be evacuated in 30 minutes if Alappad is struck by another disaster.

Beiträge zur Pädagogischen und Rehabilitationspsychologie

Herausgegeben von Evelin Witruk

Band 1 Evelin Witruk / David Riha / Alexandra Teichert / Norman Haase / Marcus Stueck (eds.): Learning, Adjustment and Stress Disorders. With Special Reference to Tsunami Affected Regions. 2010.

Band 2 Ulrike Quast: Lernermerkmale, Lernertypen, Lernverhalten. Aspekte der differentiellen Lernpsychologie für Lehrende und Lernende. 2011.

Band 3 Sabine Schneider: Synästhesie – Nachweis der Stabilität alphanumerischer Farbsynästhesien. 2011.

Band 4 Evelin Witruk / Arndt Wilcke (eds.): Historical and Cross-Cultural Aspects of Psychology. 2013.

Band 5 Johanna Sophie von Lieres: Tsunami in Kerala, India: Long-Term Psychological Distress, Sense of Coherence, Social Support, and Coping in a Non-Industrialized Setting. 2013.

www.peterlang.de

FSC
www.fsc.org
MIX
Papier | Fördert gute Waldnutzung
FSC® C083411

Kolibri 360 GmbH
Ferdinand-Jühlke-Straße 7
99095 Erfurt, Deutschland
produktsicherheit@kolibri360.de